Fitface Foundations
Face Exercises

Charlotte Hamilton

Published in the USA & United Kingdom
By
Fitface

Copyright © 2008 Charlotte Hamilton

For bulk order purchases or for other information please visit
www.fitfacetoning.com & products on www.fitface.co

Typeset, printed and bound in the USA & United Kingdom

By the same author

In the same series
Fitface - Hands Free Facial Toning Exercises 2010 & 2011
Fitface - The Guide to Fun Facial Toning Exercises 2008

Under the pseudonym
Colette Sinclair

MAN HUNT

First published:
Sidgwick and Jackson - Pan Macmillan Press 1989

For my daughter Brittany whom I love madly

and

Nami wherever he may be.

With special thanks to:

Captain Peter: he listened and encouraged me to finish.
Coco: she inspired me to continue to tell the world,
"There is another way to look beautiful forever."
Mummie and Brian Blunden: for editing.
They all made this book possible.

About the Author

Born and raised in England. Charlotte immigrated to California USA in her twenties. There she witnessed firsthand the butchery of a major face-lift to her former partner and vowed never to have a face-lift. Following the breakup of her marriage she returned to England to raise her new baby daughter. Eighteen years later, her daughter suffered a horrendous near fatal accident in which she sustained major facial trauma with 21 fractures to her face. She survived, graduated with an honours' degree and now lives in Australia. Charlotte returned to live in Florida, USA although currently she resides on the south coast of England.

Charlotte loves to hear from her readers and will endeavour to answer every single e-mail. She invites you to visit her website at www.Fitfacetoning.com and coming soon www.Fitface.co.

Contents

FACE EXERCISES

With photographs showing you how to exercise

Chapter 1

The Foundations

The Fitface Toning Rules

Essential reading!

<u>**Do not perform these exercises**</u> if you have ever had any facial surgery, have used any Botox, derma fillers or injectables without first consulting a dermatologist or surgeon (as applicable). **Fitface toning** exercises are preventative measures to avoid or postpone invasive procedures.

1. Do not perform **Fitface toning** exercises for at least 6 months if you have had a laser or chemical peel as your skin will be fragile. Consult with your dermatologist or plastic surgeon as to precisely how many weeks or months it will be until your skin is fully healed.
2. Do not exercise if you have sunburn, the skin will be tight and fragile. Wait until the natural oils and body's own hydration returns to your skin.
3. Do not exercise if you are on pain killers as these can mask any of the warnings signs to stop exercising.
4. **Stop** exercising immediately if you feel any pain or twinges.
5. **Stop** exercising if your muscles start to quiver. It is a sign that you are doing too much. Relax, there is always tomorrow.
6. Do not exercise your facial muscles if you feel tired, emotionally upset or if you are hung over because you won't enjoy it and subconsciously you will remember the negative experience. You can change brain plasticity either negatively or positively. Let the brain record the positive repetitive changes only!
7. Be especially cautious with the neck exercises, take it very gently and **totally avoid them** if you experience any discomfort, aches or back problems. Don't bring your head too far back, if it says "tilt your head", that's what it means. It does not mean to thrust your head as far back as humanly possible!
8. When beginning exercising use a mirror initially, unless specifically stated not to do so. When you become proficient at each of the exercises you will no longer need to look in the

mirror, except to ensure that you are continually doing them correctly.

9. Start slowly; get used to each of the exercises. If you find one too difficult at first, leave it out and come back to it later. Your muscle tone will improve over time and therefore, as with any exercise you will be able to do more, eventually. Remember exercising must be fun as well as good for you.

10. Remember that frequency is more important than the length of time spent on **Fitface toning**. Begin with a minimum of 15 minutes every day for the first three months. After that continue with a maintenance programme of 3 to 4 times a week.

11. Remember less is more; do not do more than half an hour of continuous **Fitface toning** exercises in any one day. However, be sensible; when you are learning, understand that what should take 20 minutes may take an hour!

12. Begin gently, if you find either the number of repetitions or the length of the hold time too difficult at first, relax. Move on. Over time, you may gradually work up to the required holding time or number of repetitions. If not, it is not the end of the world, just enjoy your routine.

13. Always relax your shoulders.

14. Always remember to breathe.

15. Maintain a healthy well rounded programme adding in a couple of extra exercises for problem areas in addition to the basic routine. Just as you don't work exclusively on one muscle at the gym, you work the whole body; work your whole face with **Fitface toning.** The exercises are designed to be mixed and matched. **Do not concentrate on just your problem areas or only a few exercises. IT WON'T WORK and this is the biggest mistake most people make.**

16. Eventually learn a **Fitface toning** routine by heart so that you can perform it anywhere, even watching television or in the tub (bath)!

17. Remember this for later. If on the first day, or the second day, or whenever you try **Fitface toning** and your muscles become sore, **stop**. Do not over do it. Come back. Try again. Start over. It is only natural that when you move muscles that have not been used much before, that once stretched they will feel sore. Soon they will rebuild and become stronger, tighter than ever before.

Getting started with Fitface facial toning exercises

What will you need most?
Your ambition!

Change

For most of us starting something new is both scary and exciting. We all react differently to change. Some people fail at the first anxious thought of change, resist and never even get started. Others try with a running start but fall at the first hurdle and just give up! Then there are those who rise to the challenge, grasp hold of their mindset and approach change with vigour embracing the possibilities! The majority of people's attitude to change is to fall somewhere between the two extremes. The only way I can support you through your learning curve is to state that <u>if you don't try, you are guaranteed to fail</u>. Remember this and you are halfway there.

Self discipline

The only way that the toning exercises will work for you is if you do them regularly. You must also enjoy them. If you make them fun and easy, they will be! Plan to do them every day for the first three months and then 3 to 4 times per week. Sticking to a routine is always difficult but rewarding. You will not only look better physically but mentally, you will feel better for the achievement. Pat yourself on the back.

Belief

You must believe that the time, commitment, dedication and effort will pay off. Working out doesn't stop at the neck. **Fitface toning** works. Reread the pages on all the benefits of facial toning or just think about the possible horrors of surgery.

Motivation

Motivation is the **<u>single most important key ingredient</u>** of the successful outcome of any new venture you undertake. Encourage yourself, tell yourself you "love" doing **Fitface toning** exercises and you will. Plan to succeed and you will. Teach your brain to think the right thoughts; you are what you think you are. If you "think" you will, you will, if you think you can't you won't. It's never too late to start. I have included a few inspirational quotes on a separate page for you to turn to when motivation is needed.

Commitment
Facial exercises are not a quick fix. Immediate change in your appearance is not going to be dramatic. Muscular development doesn't happen over night, it is a process, it will take time; please don't give up at the first hurdle. The result of your time, dedication and commitment will have lasting effects.

Dedication
You will need to take **15 or 20 minutes out of everyday** to do these exercises for the first 3 months, you will and can find the time if you really want to. Maybe try to fit in some toning while you watch a TV programme or wait for the washing to dry, whatever, if you want to find the time you will! Later, why not incorporate **Fitface toning** into your work out routine or do facial exercises on alternate days instead of bodywork on your 'off days'.

Concentration
You can learn to move any muscle, it just takes dedication and concentration, tell yourself you can move that muscle and you will, believe, it takes practise. As a child were you not taught to, "Try, try and try again?"

Patience
You will eventually learn the exercises by heart and the routines will only take 15 or 20 minutes, it will not happen over night. It takes time and patience to become proficient. Remember that on the first day you will be exasperated and the routine may end up taking an hour to do just a few exercises. Don't worry, don't be concerned. Ultimately, **Fitface toning** will become second nature to you.

Understanding
Listen to your body, if you think you are doing too much you probably are. If there is any pain or discomfort stop. It's your body talking to you, **listen.** It knows best!

Other things you will need:
1. To consult with your physician if you are in any doubt whatsoever about your personal circumstances as to whether or not you should perform **Fitface toning**.
2. A clean face and neck (no make-up) freshly moisturised.
3. A glass of water.

4. A clock with a second hand.
5. Your hair pulled back away from the face.
6. A mirror, initially on a table.
7. A chair.

You are now ready to begin on improving and toning the already beautiful you. Whatever you need to do to succeed, **just do it** (as now coined by **Nike**). If you need to write these inspirational, thought provoking motivational sayings down, buy the posters, memorise these phrases and stick them in your bathroom, do whatever it takes.

Inspirational quotes

My personal favourite is my school motto

Beyond the best, there is a better!

Winners

*While most are dreaming of success, winners, wake-up
and work hard to achieve it.*

Only number 1 is remembered, don't settle for second place.

Success

*You have to think you can win.
You have to feel you can win.
You have to believe you can win.
You have to know you can win.
You have to be certain - that winning is the only option.*

Goals

*Have one only.
Be single minded in your approach.*

Challenge

*Relish the thought of new experiences to stretch yourself
to your full potential.
Get out of your comfort zone!*

Commitment

When the going gets tough, the tough get going.

Sorts the winners, from the losers.

Perseverance

The difference between a successful person and others:
it is not, a lack of strength,
it is not, a lack of knowledge
but is, rather a lack of will.

Determination

I will do it.
Failure is not an option.

Fitface toning training programmes

Basic

The **Fitface toning** training programme begins with a basic introductory easy **Fitface toning** programme. This programme is what it says it is a **Basic,** no nonsense anytime, everywhere routine that is suitable for anyone. The basic programme can done by anyone without any training or learning. The other programmes require more effort and need to be learnt and practised.

There are three different **Fitface toning** training programmes, beginning with the **Anytime Fitface** toning programme at the beginner's level, advancing through to the **Standard Fitface** toning programme and finishing with the **Advanced Fitface** toning routine. A 3 step programme, built for you to gently work through at your own pace. Each programme is designed for a different stage of your personal development, with 3 separate and distinct levels within each programme to progress through.

Additionally, there is a **Micro Fitface toning** workout for when time is very short which only takes 3 minutes.

Programmes

Step 1 Anytime   Fitface Programme
Step 2 Standard   Fitface Programme
Step 3 Advanced   Fitface Programme

+ Micro Fitface toning
Within the 3 step programme, there are 3 separate levels of attainment possible within each routine.

Levels

1. **Beginner's level**
2. **Regular level**
3. **Target level**

   +Extra Exercises only available at the Advanced Level

   - **Beginner's level** is the basic, entry level at which to start performing each of the new programmes.
   - **Regular level** is the next or the second level of expertise, for each programme.
   - **Target level** is the final upper, third level of attainment to targeted for your own unique specific facial areas that you wish to improve.
   - **Extra Exercises** are at the Advanced Level and only for the determined and dedicated.

Target

Each exercise has a target zone; this is the area of the face that the exercise is primarily working on. However, do remember that the muscles of the face all work in conjunction with one another, so you exercise one, you exercise them all. The target is only a guide to help you to target your problem areas. I stress, do not target just a few exercises, Fitface is not designed to work like that. If you strengthen one muscle exclusively another will suffer and that is, in my opinion how the fascia literately gets stuck in a groove as explained fully in **Chapter 2**.

Be patient with yourself

The brain is a remarkable organ, it has the ability to make new connections every second, constantly evolving and altering due to its plasticity and memory. Because of this, you have the power to change at any time. The more repetition an action, the more proficient you will become at it, just like riding a bike or learning a dance routine, so it is with **Fitface toning**.

At first it may be difficult, your muscles may need to strengthen in order to do some of the exercises and in time they will, but don't give up, try, try and try again. As those new connections in the brain are made, the evolution occurs and in time, the various exercises will become second nature to you as your muscles learn what you require of them. The most important thing to remember is that learning any new routine takes time until you become totally familiar with it. Please don't get frustrated with any of the **Fitface facial toning exercises**, if you can't master a pose, move on and come back to it another day. You must enjoy what you're doing.

In the beginning it will take you much, much longer to do any of the programmes. You will be looking at the book, looking at yourself, trying to master your performance and it will seem to take ages, but eventually you won't even need a mirror, and then only the **Fitface toning** guide to refresh your memory. Just remember, although you might not be able to do an exercise today, tomorrow you will. It just takes time. Practise makes perfect.

Photos

I encourage everyone to take "before and after" **Fitface** photographs. However, don't expect miracles, those only happen when the photos are touched up professionally or the lighting has been changed deliberately for celebrities or advertisements. Be realistic. Take one at the start, another about 4 months later, another at 7 months and finally a photo at a year. The photos should be taken in the same place, at about the same time, with the same lighting conditions. I would love to see the results. Do bear in mind that a year later, without **Fitface toning** you would normally have looked older, you will be amazed at the results.

Beginning

Although facial muscles work as a team, each of the twenty-six voluntary muscles of the face should be trained separately. The facial exercise instructions start with a warm up and then flow down the face, from the forehead to the neck. This order is important, not critical but I suggest you follow it.

Maintenance

For maximum results, I would recommend that you **exercise 5 times weekly or daily,** (with weekends off) **for the first three months** and **then, follow a maintenance** routine of **every other day,** alternating between <u>all</u> the different routines. However, be sensible, if your muscles are tired skip a day or two. Only do what you can. Remember that pain is a sign to stop as well as quivering muscles.

Experience has shown me and my clients that it is better to learn one programme thoroughly (in order that your muscles learn what is required of them) before even attempting to progress to the next programme or level. But even when you have completed the last programme at the advanced level, I would still suggest that you keep mixing up your programmes (depending on your mood) to give yourself a totally balanced look.

Mix and Match (IMPORTANT)

Personally, I tend to do, what I feel, when I feel like it For example; I do the **Advanced Fitface toning programme** when I'm feeling brilliant and energetic, the **Anytime Fitface toning programme** when I'm on holiday, travelling etc (because it's the easiest to remember) and the **Standard Fitface toning programme** most of the time. I also throw in some of the **Basic Fitface exercises** whenever I can. It's all a matter of choice. It is your choice, the most important thing is just to enjoy. DON'T FEEL GUILTY IF YOU MISS A DAY, a week, or even a month. Begin again. Try, try and try again, don't ever just give up for good! On the other hand don't keep procrastinating that you will start tomorrow. **Start today.**

The Fitface 3 steps programme

Commence with the **Basic Fitface**. It's fundamental, the foundations, it is also fun, it's easy and it's a start. When you are familiar with the exercises (a minimum of a week, or having completed all the exercises easily at least five times) and you want to move on to the real programme, begin with the real programme. If not, no worries, you can stay with Basic **Fitface** forever, diving in and out whenever you like, some facial exercise is better than no facial exercise.

Step 1
- **Anytime Fitface toning programme**

This routine was designed to be done anywhere, (watching TV, listening to the radio, in the tub (bath), in the dark as a passenger in a car/train/coach/plane journey) at anytime.

Eventually, you will be able to perform the programme without the use of a mirror. It's simple and super effective. This programme should be **practised for a minimum of a month** before continuing to the next programme, ideally at least six weeks.

Step 2
- **Standard Fitface toning programme**

This routine was designed to be the next stage up from the **Anytime Fitface toning programme** and therefore should not be attempted until the Regular level of the **Anytime Fitface toning programme** has been achieved and maintained for six weeks.

The previous routine has allowed your muscles to learn, build and become more familiar with some of the **Fitface toning** exercises before introducing more complicated movements. **Maintain for 6 weeks** after which you may want to progress to the next programme, but do what you can. Stay at this level for life if you prefer.

Step 3
- **Advanced Fitface toning programme**

This routine was designed to be the next step up from the **Standard Fitface toning programme** and therefore should not be attempted until the Regular level has been achieved and **maintained for more than six weeks.** The previous programme has enabled your muscles to become familiar with complex movements and promoted new muscles growth. Eventually you will be able to master the Advanced programme at the Standard level and if you wish also Target specific areas.

The **Fitface toning** programmes routines should all be used in conjunction with one another to give your facial muscles balance. The idea of the 3 step **Fitface** facial exercise routine is to give your face tone and build muscle mass.

That's it!

Exercise professionals or fanatics (only for the toughest - not me)!
If you are a professional one of those super fit young ladies (that I see at the gym 24/7) and would like to take **Fitface toning** to the max, in the shortest amount of time, then I would suggest that you learn each programme for only a month. Then continue with a 5 times a week maintenance routine at the Target level, a routine in which every other workout is the **Advanced Fitface toning programme.**

Problem areas
You may want to put extra toning emphasis on an area or areas that concerns you. You can easily accomplish this by choosing up to 3 extra exercises that target your problem area and incorporate them into any of the programmes, **see the Target level of each routine.** However, remember to work out at the less intense levels initially and to refrain from spot building (targeting an area exclusively). Do not add more than 5 minutes a day by targeting an area.

Advanced optional extras
These **Extra Exercises** were designed for those persons who have reached the **Target Level of the Advanced Programme** and STILL want additional choices with their problem areas.

Micro Fitface toning programme

When you have no time at all, or you are on holiday, give yourself a 3 minute micro work out every other day at the bare minimum. Everyday would be even better.

This is the most important exercise in the whole of the **Fitface toning programmes** and therefore it is present within each one. It is an isometric exercise, that is very difficult to learn but once mastered, this toning exercise can be performed at anytime, anywhere.

The first exercise in the **BASIC FITFACE is the "Face-lift".**

1. With eyes open (easier) or shut
2. Lift eyebrows
3. Contract muscle of scalp and temples, lifting your eyebrows up and back to connect with the back of the neck

   - Hold for 1 minute
   - Repeat twice (i.e. for a total of 3 times/minutes)
   -

Note: Feel the whole face pull taut, up and back. Imagine that your whole skull is being tightened, lifted and pulled back.

Tip: Ensure that the sides of your mouth are level with the corners slightly turning up. **DO NOT FROWN**

TARGET Whole face, especially forehead and cheekbones

Daily life tips for a firmer face
- Relax when you are thinking, don't permanently frown
- Hold an exaggerated smile for a minute or two every day. Not only will this make you look better but you will feel better too!
- Once you have totally learned a **Fitface** routine you have the ideal opportunity to "multi task", try doing some isometric exercises at the same time. While doing some of the easier **Fitface toning** exercises try to tighten your abs, legs or buttocks and pull your shoulders back all at the same time!
- Once you become proficient at the toning exercise it's also a great time to tone those internal muscles

- Remember, everyone progresses at a different rate. An exercise that is difficult for one person might be easy for another, accept this, it is just the way we are all made slightly differently and **Fitface** is not a competition. It should only be performed to make you feel better about yourself.

Programmes duration

This depends totally on how familiar you are with the exercises. In broad terms I think of each exercise as taking approximately 1 minute. Each programme has 14/15 exercises with the **Micro** or **Face-lift** taking only 3 minutes. Therefore I would expect you to spend approximately 20 minutes on each programme.

Chapter 2
BASIC FITFACE
1 Face-lift

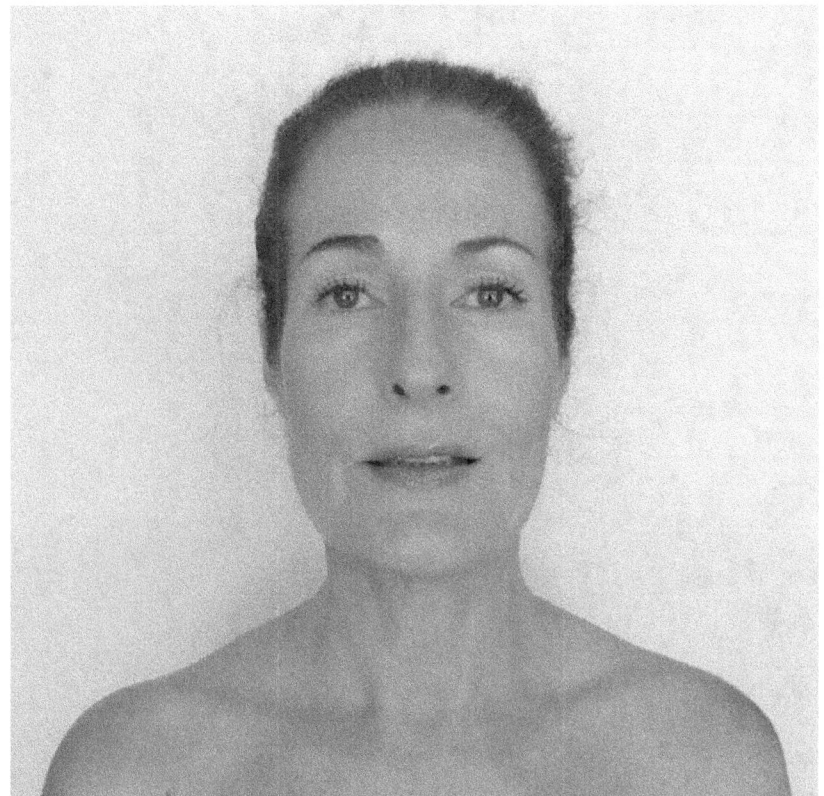

TARGET Whole face, especially forehead and cheekbones

1. With eyes open (easier) or better shut
2. Lift eyebrows
3. Contract muscle of scalp and temples, lifting you eyebrows up and back to connect with the back of the neck
- Hold for I minute
- Repeat twice (i.e. for a for a total of 3 times)

Note:
Feel the whole face pull taut, up and back. Imagine that your whole skull is being tightened, lifted and pulled back.
Tip:
Ensure that the sides of your mouth are level with the corners slightly turning up. **DO NOT FROWN**

BASIC
2 Kiss

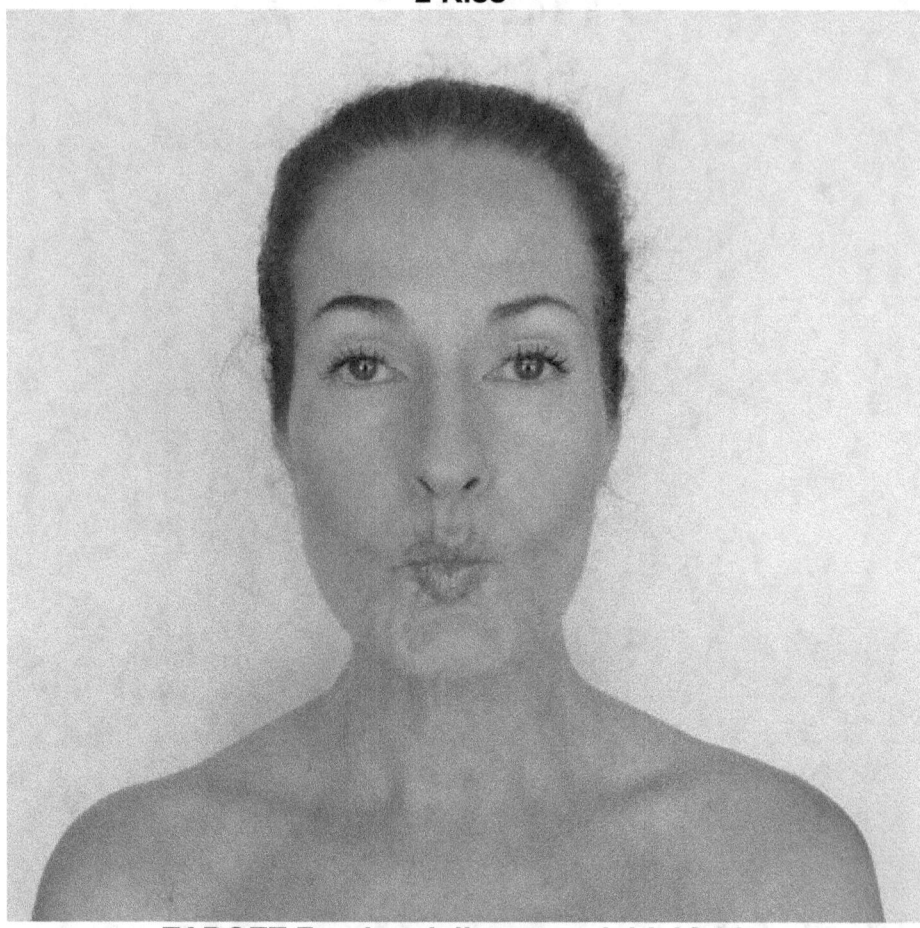

TARGET Forehead, lips, *nasolabial* folds

1. With closed lips, move lips forward into a kiss
2. Lift eyebrows high

- Hold for 1 minute

BASIC
3 Puffers

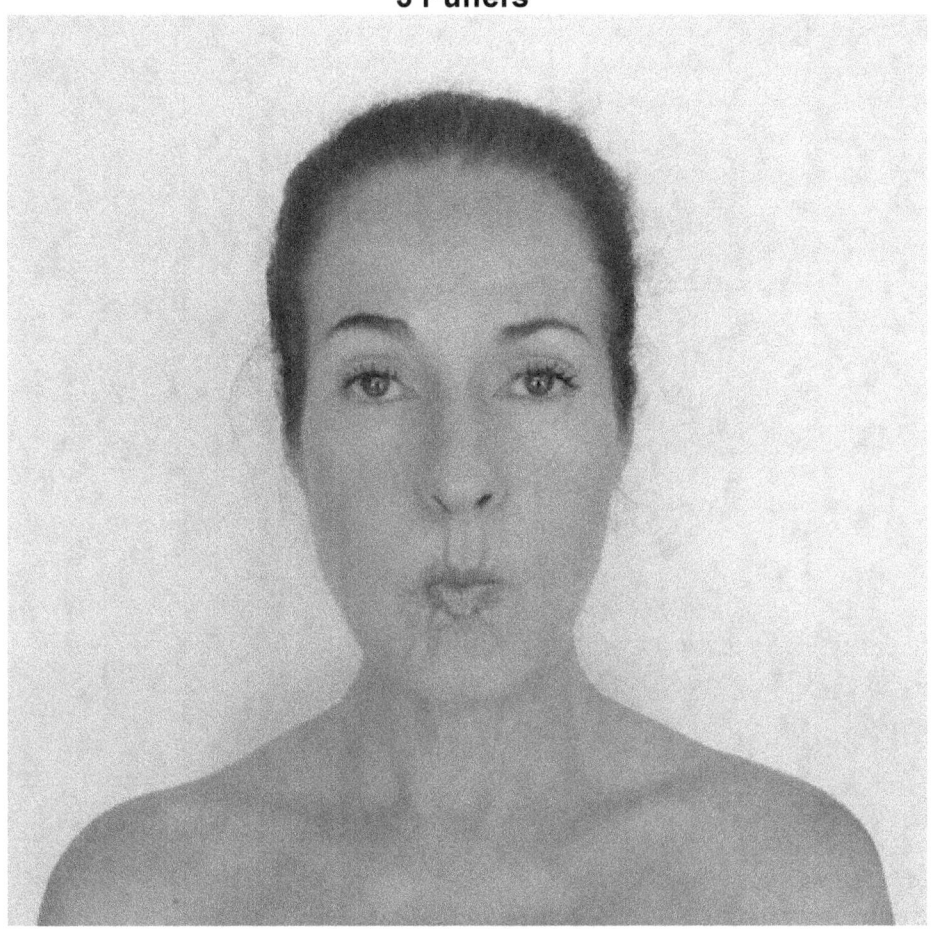

TARGET Cheeks

1. Blow up cheeks with air
2. Lift eyebrows (if you want an extra pull)

- Hold for 1 minute

BASIC
4 Grimaces

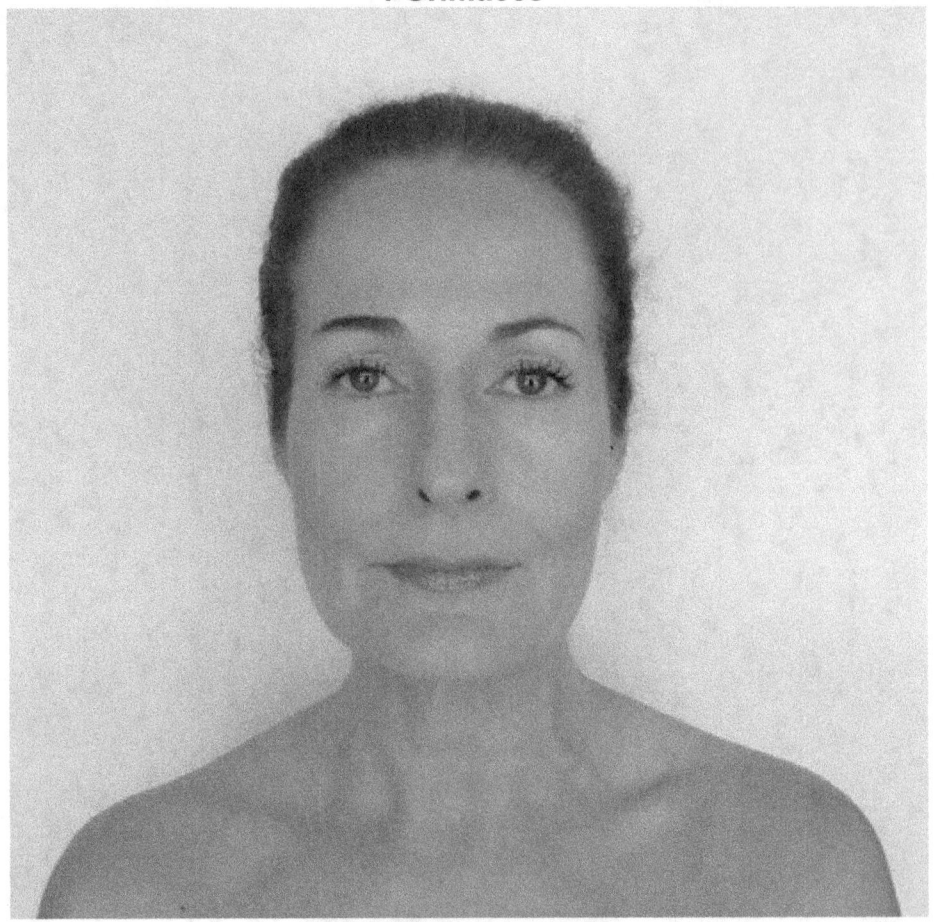

TARGET Pull up jowls

1. Lips closed, make a broad straight smile
2. Lift eyebrows (if you want an extra pull)

- Hold for 1 minute

BASIC
5 Happy smiles

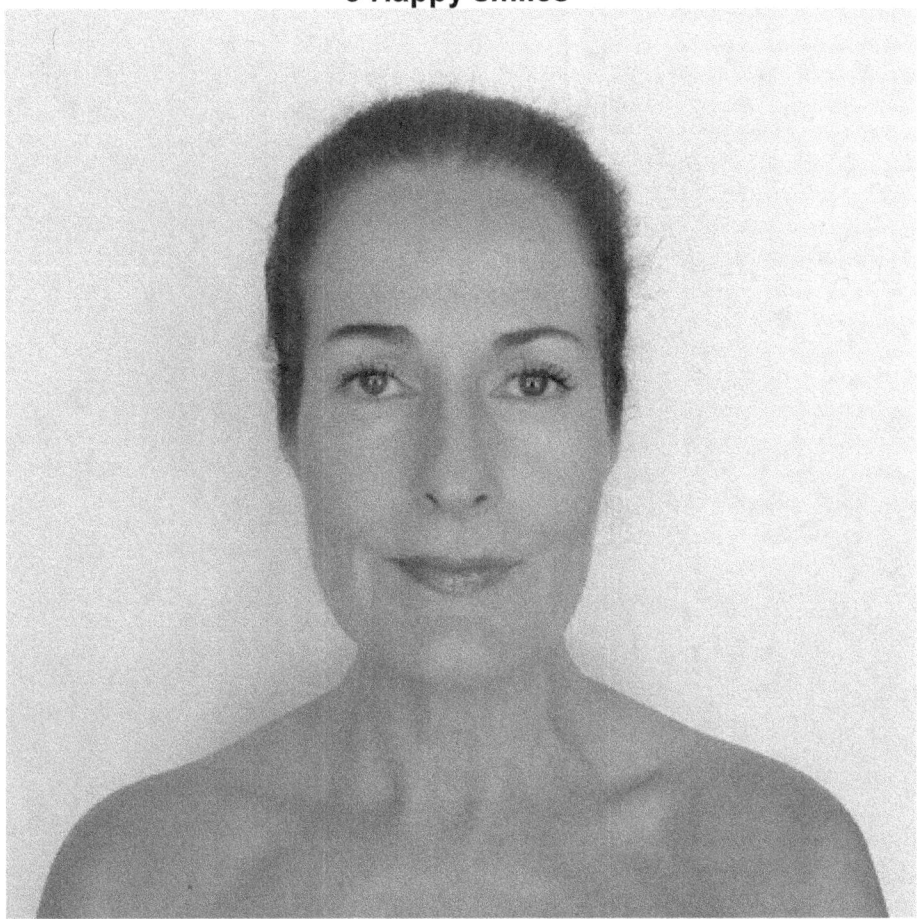

TARGET Pull up jowls, lift lower face

1. Lips closed, make a broad smile and extend to the earlobes
2. Lift eyebrows (if you want an extra pull)

- Hold for 1 minute

BASIC
6 Laughs

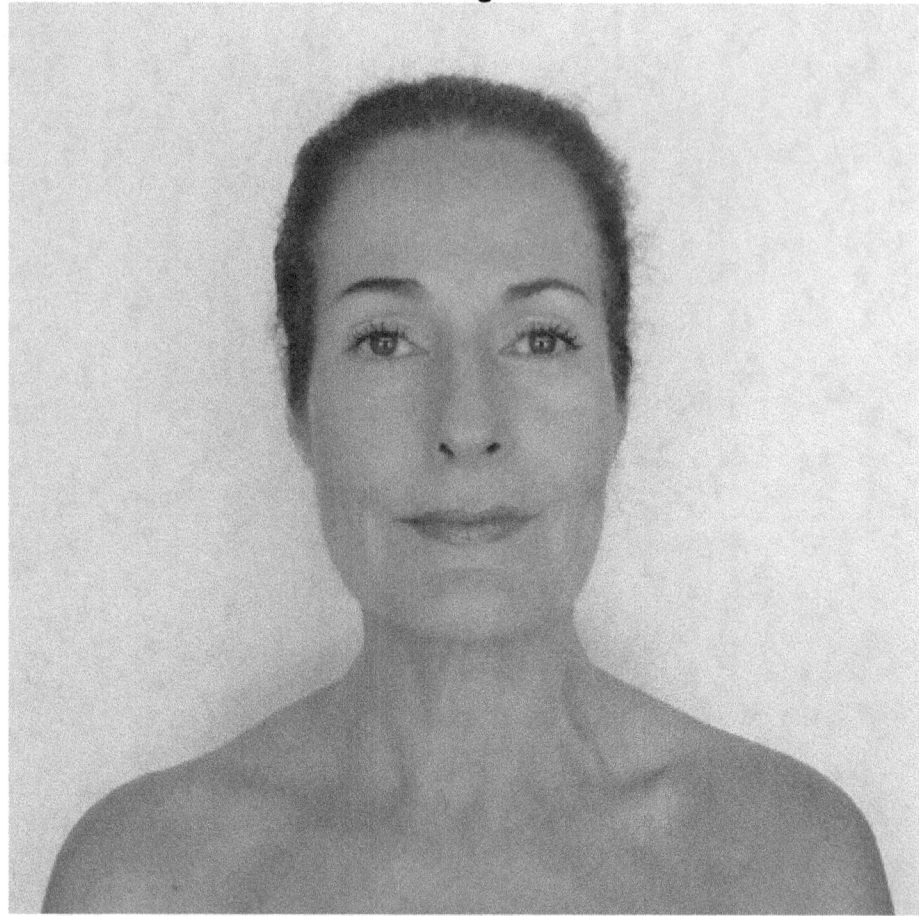

TARGET Jowls and face-lift and under eye

1. Lips closed, make a big broad smile, huge smile right up to the corners of the eyes – yes, showing crow's feet!
2. Lift eyebrows (if you want an extra pull)

- Hold for 1 minute

Note:
Your lips will want to part but don't let them.

BASIC
7 Snarls

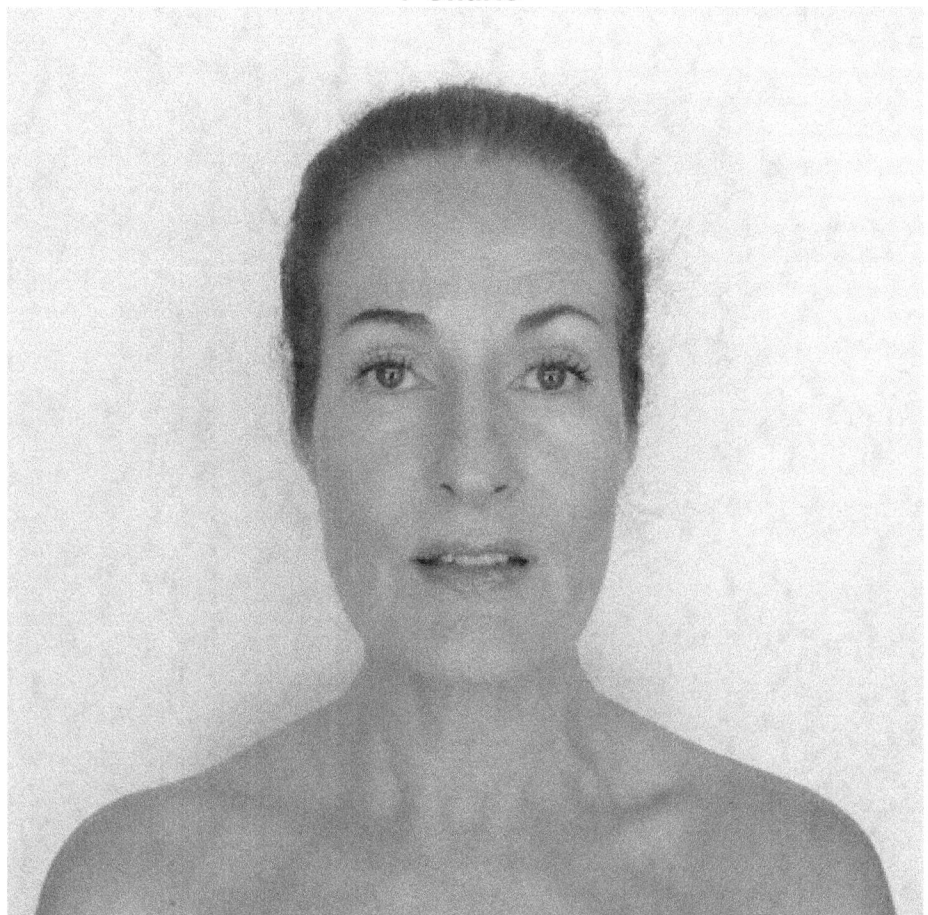

TARGET Cheeks and nasolabial folds

1. Begin with lips closed and raise your top lip only avoid the bottom lips natural tendency to pull downward, instead try to keep the bottom lip straight)
2. Lift eyebrows (if you want an extra pull)

- Hold for 1 minute

Note:
Try to feel the outer top of the snarl being pulled up by the temple.

BASIC
8 Flutters

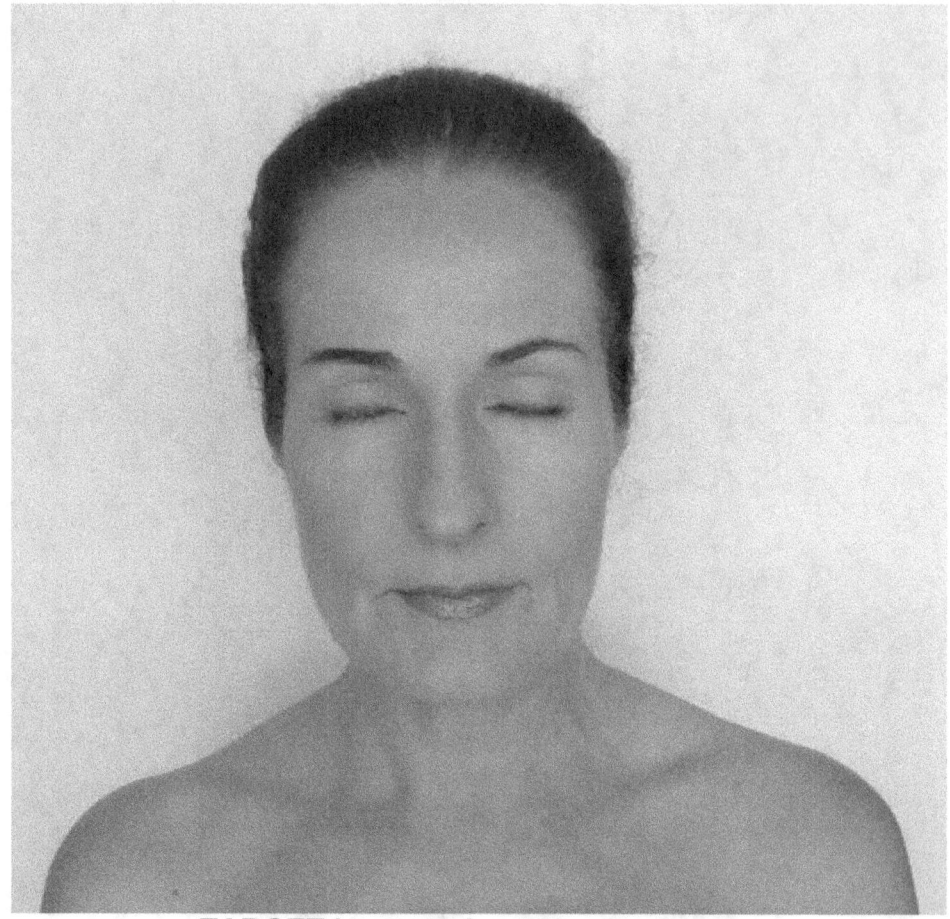

TARGET Inner and outer eye corners

1. Flutter your eyelashes as quickly as possible, completely opening and closing the eye i.e. very rapid blinks
2. Lift eyebrows (if you want an extra pull)

• Continue for 1 minute

BASIC
9 Winks

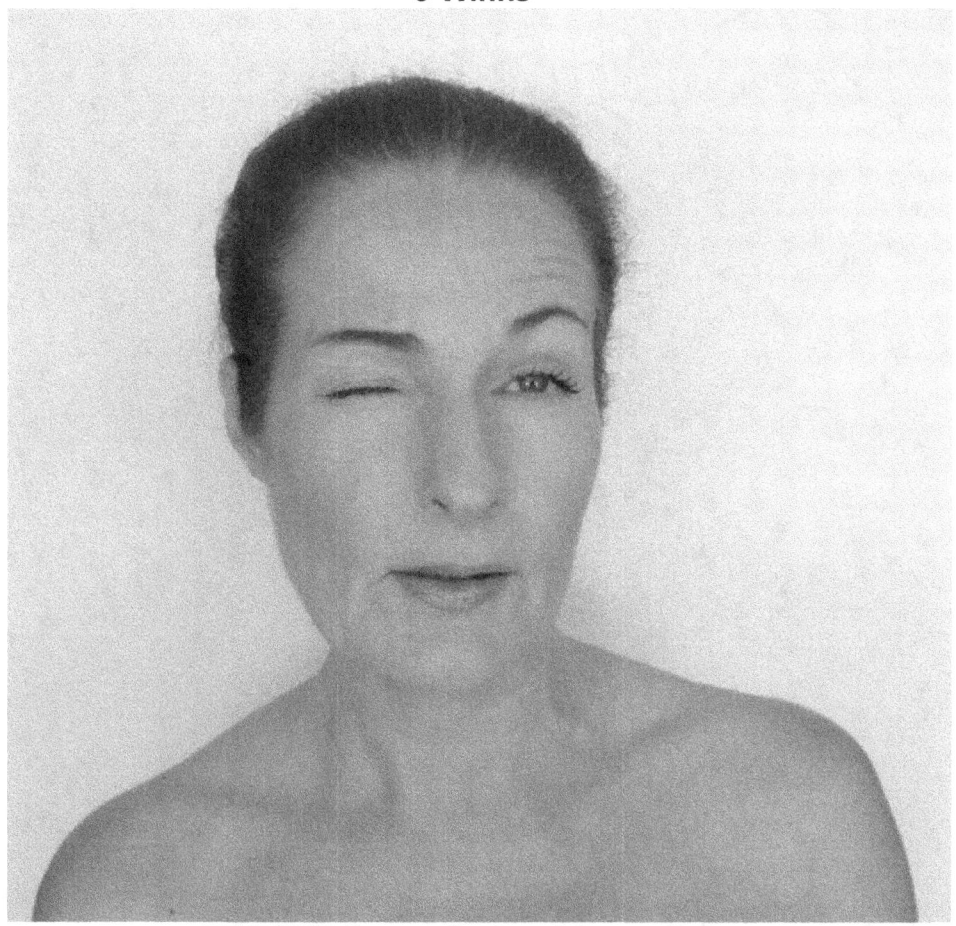

TARGET Crows feet and frown lines

1. Keep one eye open and the other shut and wink fully on one side
2. Repeat with the other eye

- 30 repetitions each eye. Approximately 1 minute total exercise

Note:
If you should find that one eye is stronger than the other, or you can wink with one eye but not the other, practise more with the weaker eye. For example do 45 repetitions on the weak eye and only 15 on the stronger eye.

BASIC
10 Stares

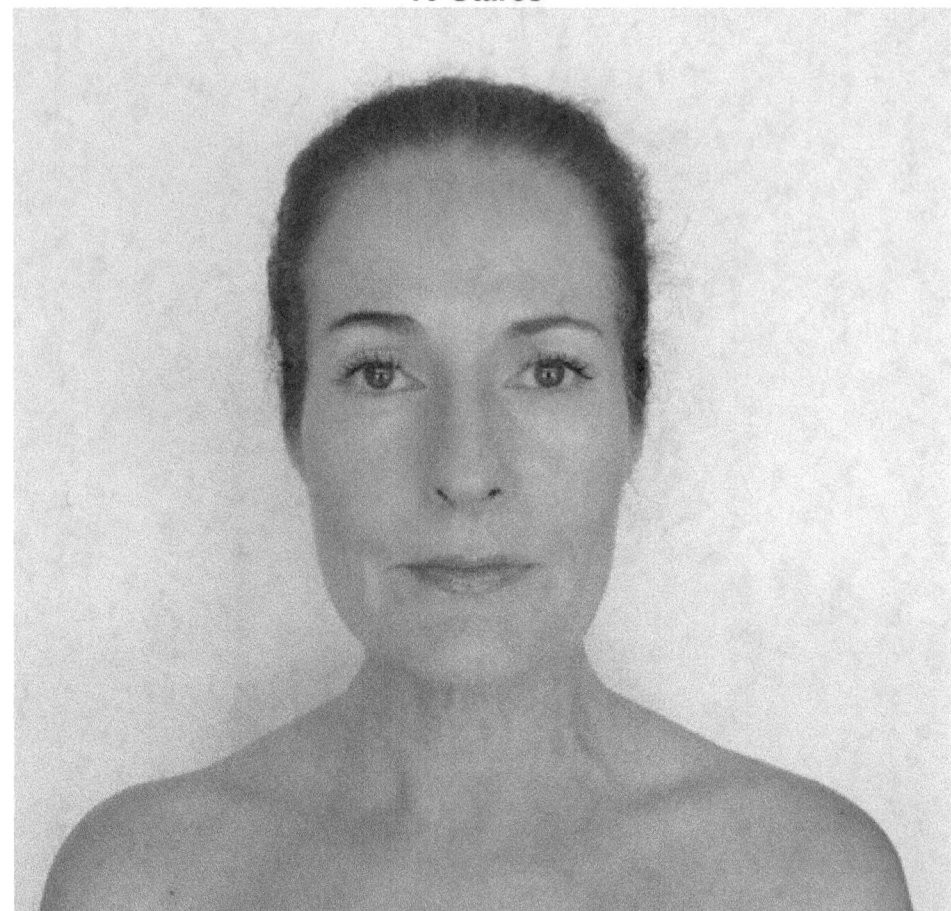

TARGET Frown lines, eye tendons and muscles around the eyes

1. Open your eyes wide and stare at a distance object
2. Hold as long as you can. Blink, when you have to and start again

• Continue doing this for approximately1minute

BASIC
11 Moons

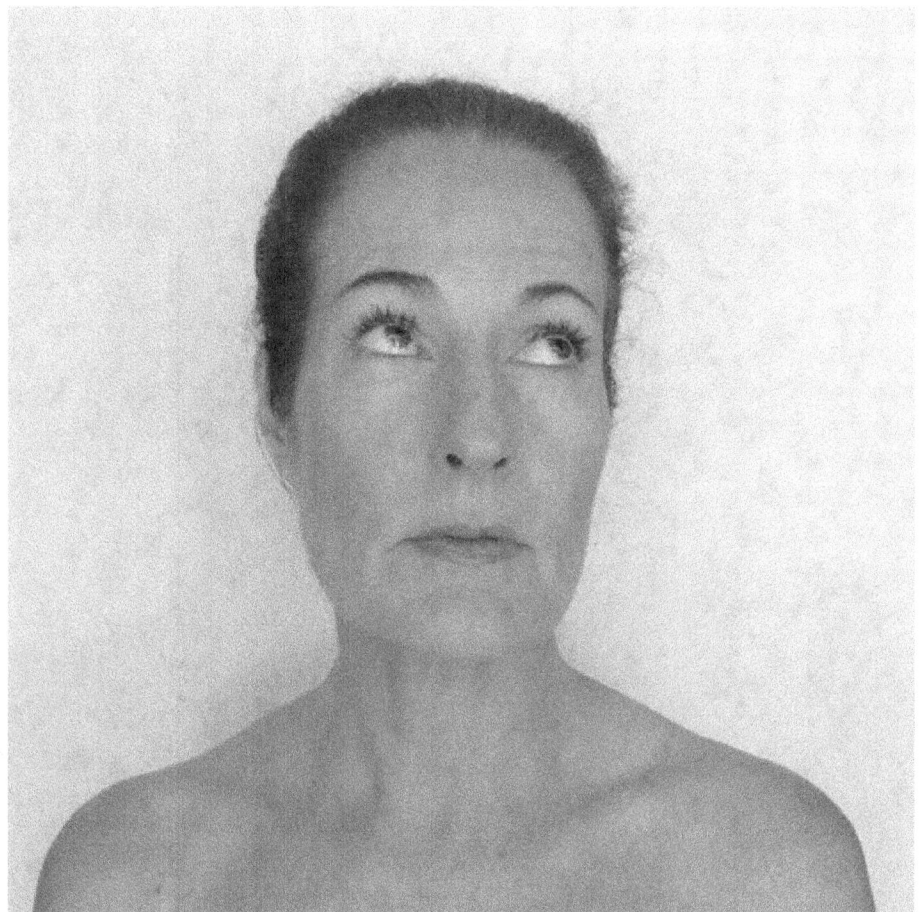

TARGET Temples, frown lines and tendons of the eye

1. Look as far as possible to one side, and circle eyes upwards looking at the tops of your eyebrows and round over and down to the opposite side
2. Lift the eyes and circle upwards to the opposite side.

*   Continue doing this for approximately 1 minute

Note:
It feels like making crescent moons repeatedly from side to side.

BASIC
12 Pouches

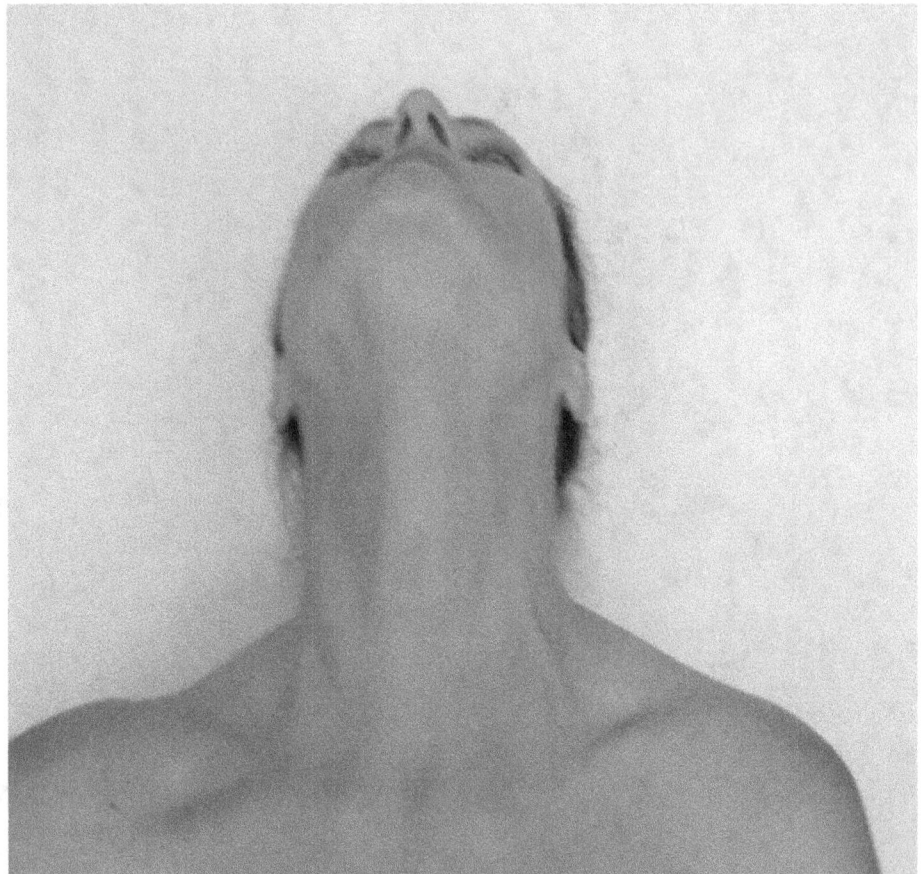

TARGET Neck, jowls & under chin

1. Ensure that your neck is long and relaxed before gently dropping your head backwards
2. Push your tongue up into the roof of your mouth.
3. Hold for approximately 20 seconds, release your tongue
4. Bring your head into the upright position

- Repeat twice

Note:
You may not be able to hold this position for 20 seconds. For example: If you can only manage a 10 second hold, just repeat the exercise four times instead of a total of 3 times.

BASIC
13 Scream

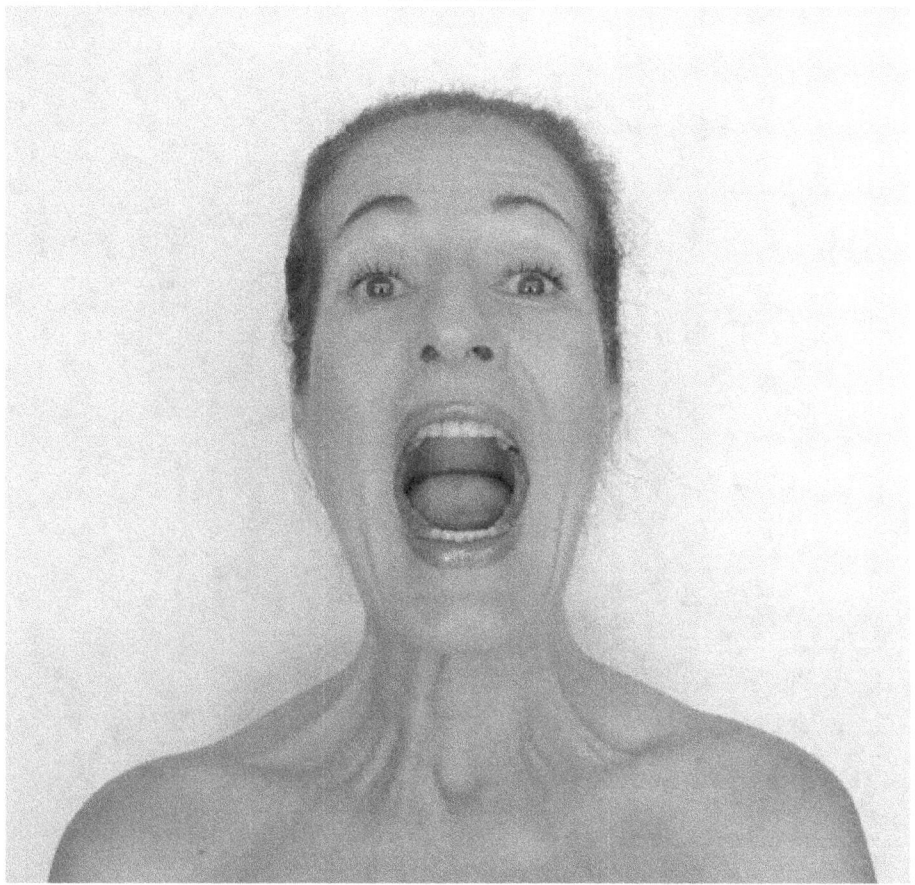

TARGET Everything, whole face.

1. Make a scream (without making a noise)

• Hold taut for 1 minute

Note:
If this is too much, hold for 30 seconds and repeat. You should feel the whole face come alive and feel as if it has been fully stretched.

BASIC
14 Umm

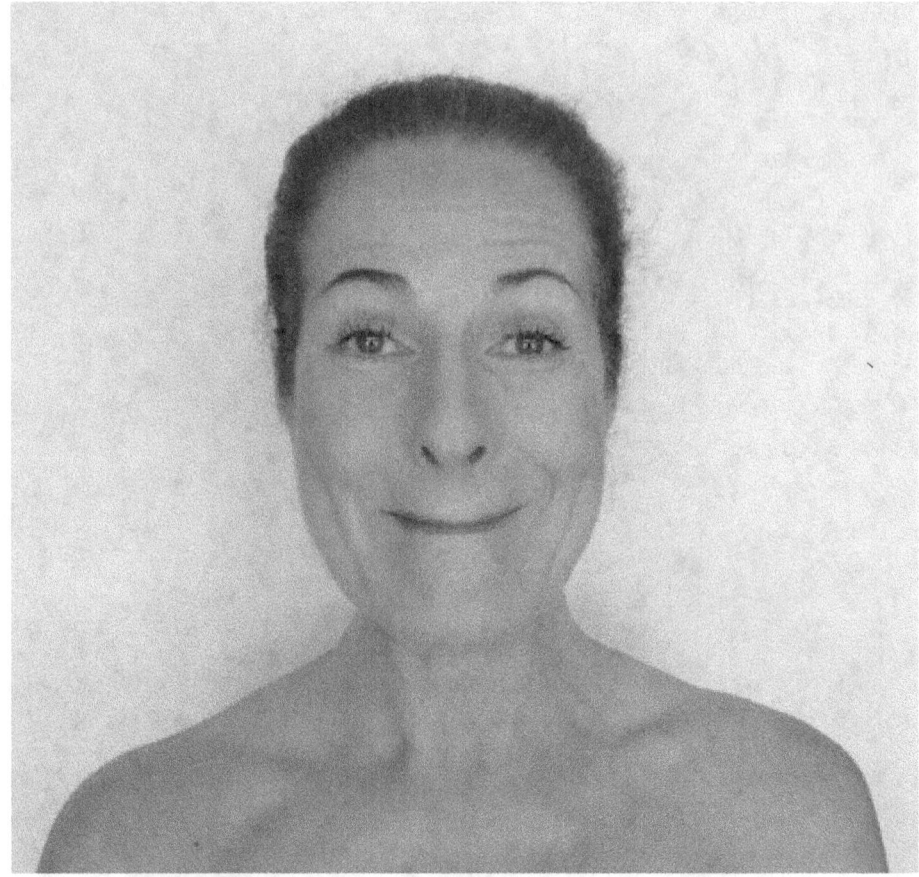

TARGET Cheeks, under eye area & eyebrows

1. Pull top lip over teeth
2. Push bottom jaw forward and pull bottom lip over teeth
3. Pull up sides of nose, cheeks

- Hold continuously for approximately 1 minute

Note:
This is difficult and may take you a while to master, start with holding for less time and work up to 1 minute.

Chapter 3

Step 1

Anytime Fitface toning programme

The **Anytime Fitface toning** programme was designed to be your introduction to **Fitface toning**, to be done anywhere (whilst watching TV, listening to the radio, in the bath) at anytime. The Anytime **Fitface toning** programme is relatively easy to learn. Once learnt you will be able to complete the programme without a mirror! It's a simple and super effective programme at the Regular level.

Do not progress to the next stage (Standard) until you have allowed your muscles to build and become familiar with some of the basic toning exercises before introducing those more complex movements. The routine should be used in conjunction with the other programmes to give your face a balanced training routine. The idea of the 3 step **Fitface toning** programme is to give your face a toned appearance.

There are three levels to progress through: Beginners, Regular and Target. The first two levels in each programme must be completed before continuing to the next programme stage. The Target level of each programme is optional; it is the Regular level but more intense with the opportunity to select additional exercises to target your own personal problem areas.

Beginner's level

Week one
Either cut the repetitions or the duration of each exercise in half.
Only when you are ready, continue to the next **Regular level.**

Regular level

Continue with the full programme for the next six weeks.
Only when you are ready, continue to the last target level of the programme or change to the Standard **Fitface toning** programme.

Target level

Select any 3 additional toning exercises from:
The Anytime, Standard or Advanced Fitface toning programmes, which you
feel would be the most beneficial for you, to target your personal problem areas.

Notes:

Timings: are for guidance only
Count: means timing in approximate seconds

Duration

This depends totally on how familiar you are with the exercises. In round terms I think of each exercise as taking approximately 1 minute. Each programme has 14/15 exercises with the **Micro** or **Face-lift** taking 3 minutes. Therefore I would expect you to spend approximately 20 minutes on each programme for maximum results.

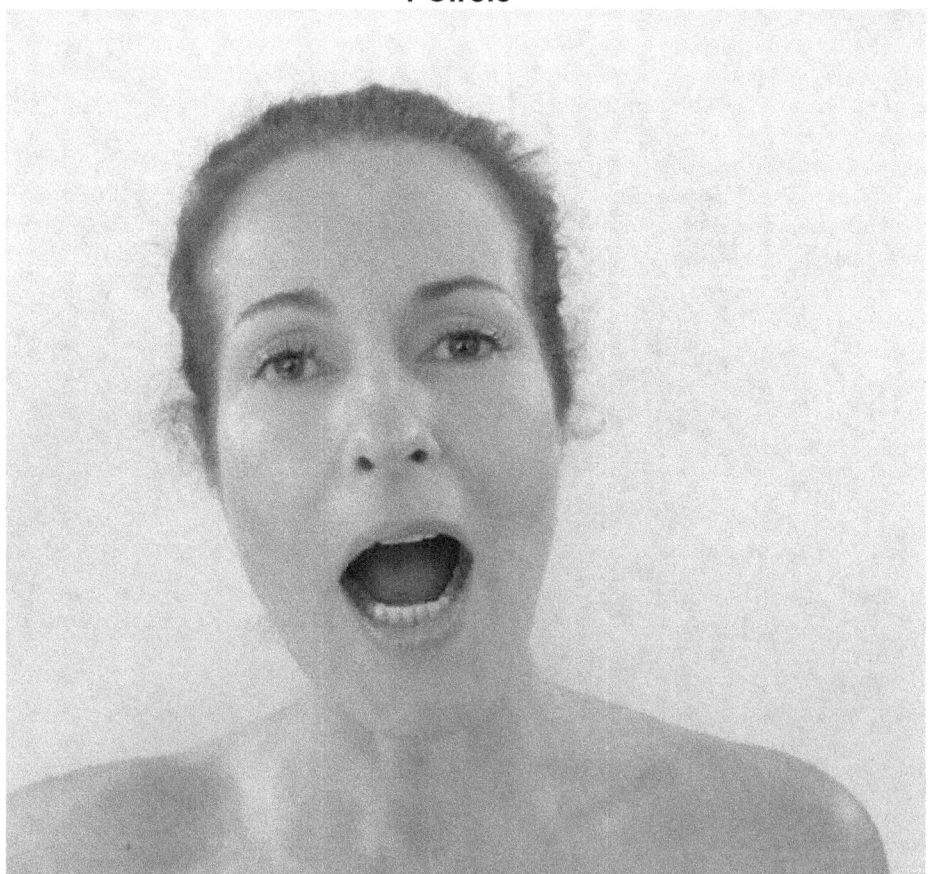

TARGET Warm up - whole face

1. Open your mouth fairly wide
2. Commence an exaggerated circular motion i.e. with your jaw forwards, then downwards, backwards, then upwards rolling forwards

- Repeat continuously for 30 rotations in one direction and then reverse, and continue with 30 movements in the other direction
- Total 1 minute

Note:
Feel as if, you are creating chewing circles forward; then reverse, backwards

ANYTIME
2 Eyebrow lifts

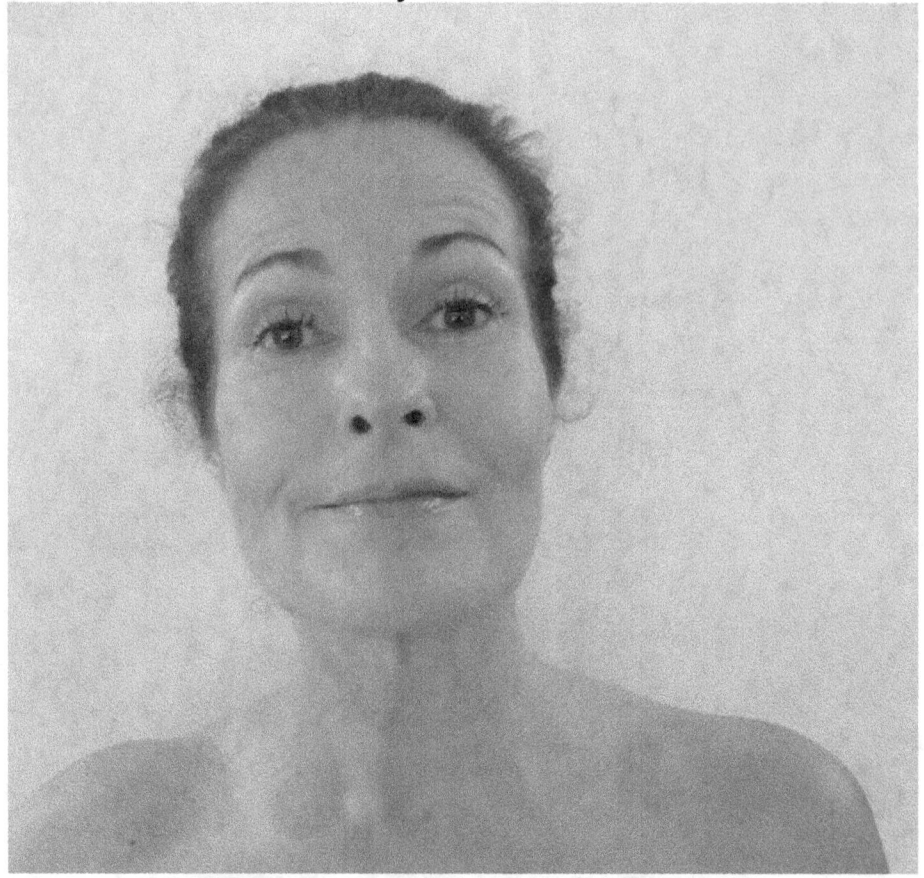

TARGET Brow, eyebrows & upper eyelid

1. Smile gently and hold
2. Lift your eyebrows as high as possible and pause for a couple of seconds
3. Relax

- Repeat approximately 30 times
- Total 1 minute

Note:
Feel as if, you are raising and lowering your eyebrows slowly.

ANYTIME
3 Eyebrow hold

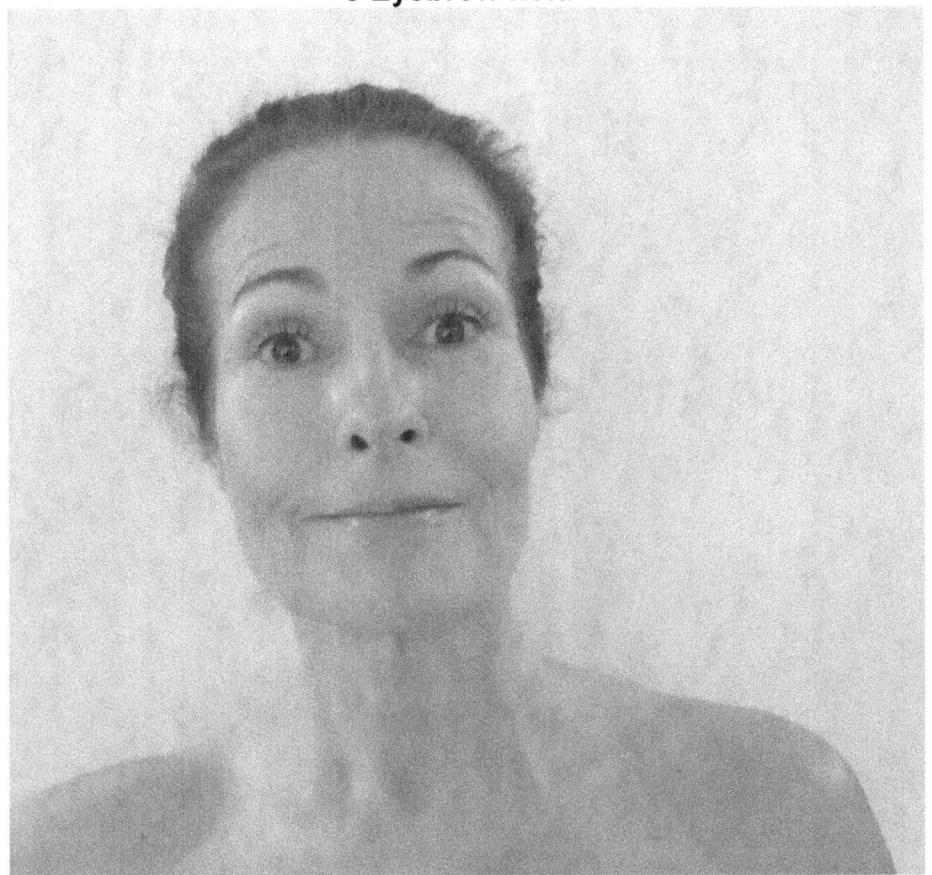

TARGET Brow, eyebrows & upper eyelid

1. Smile gently and hold
2. Lift your eyebrows as high as possible

- Hold for 1 minute

Note:
If this is too long for you, reduce the time by half and repeat twice.

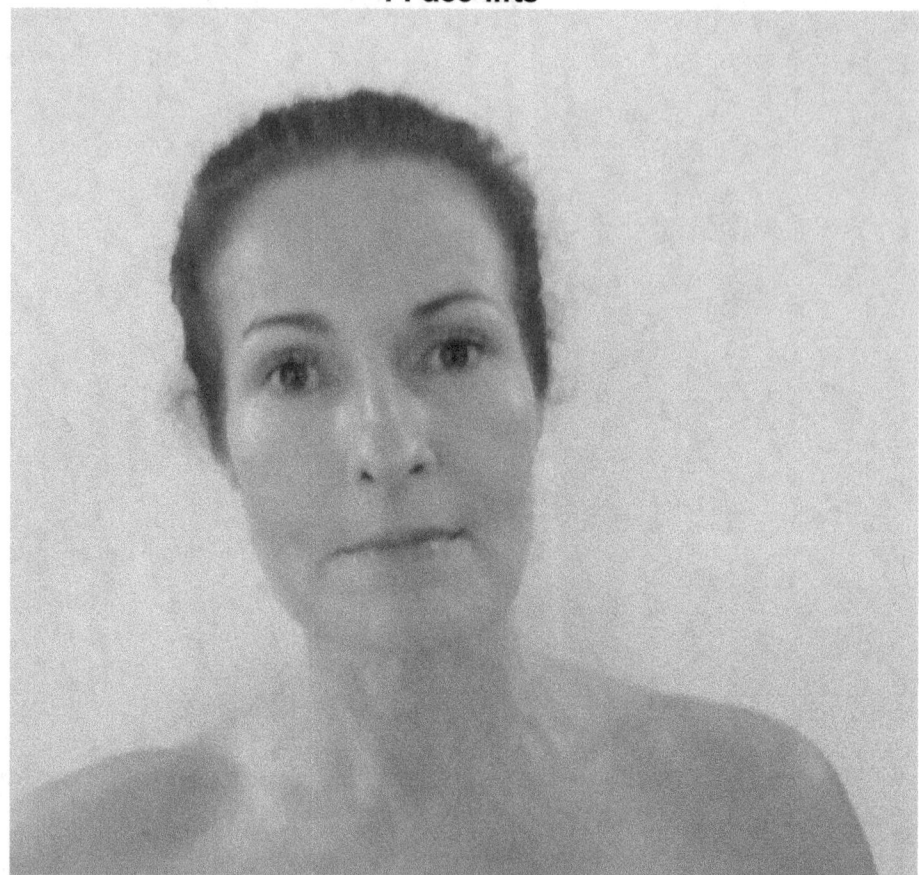

TARGET Whole face, especially forehead and cheekbones

1. With eyes open (easier) or better shut
2. Lift eyebrows
3. Contract muscle of scalp and temples, lifting you eyebrows up and back to connect with the back of the neck.
- Hold for I minute
- Repeat twice (i.e. for a for a total of 3 times)

Note:
Feel the whole face pull taut, up and back. Imagine that your whole skull is being tightened, lifted and pulled back.

Tip:
Ensure that the sides of your mouth are level with the corners slightly turning up. **DO NOT FROWN**

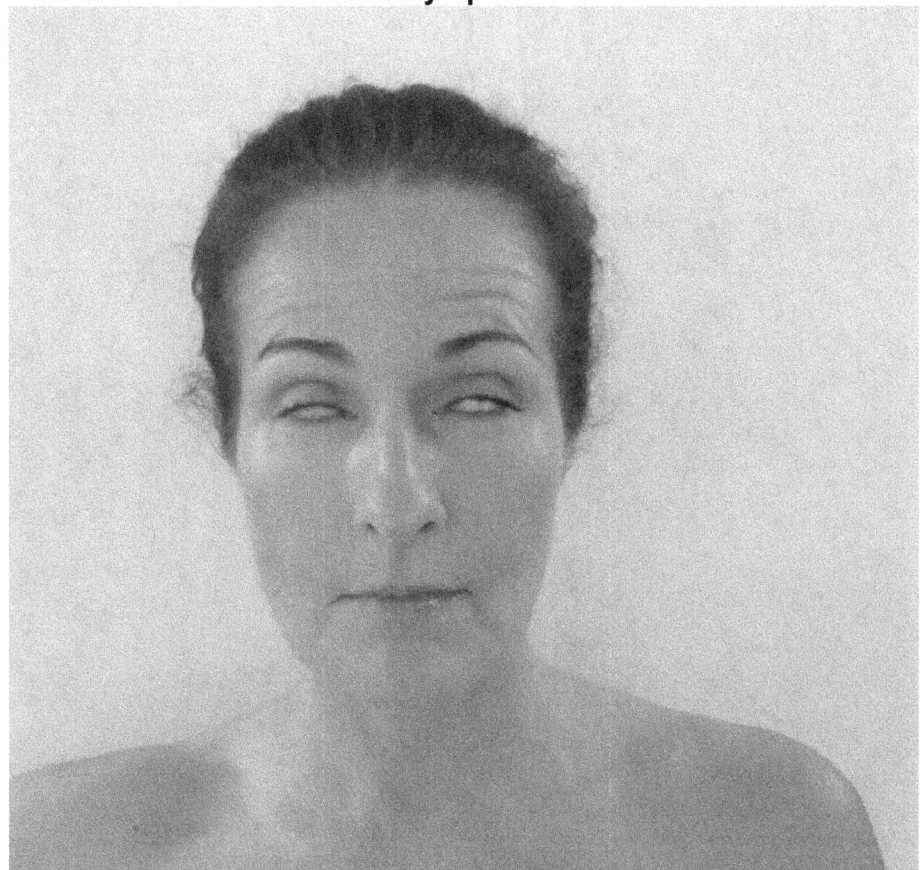

TARGET All around the eyes, both corners, plus under eyelids

1. Begin with eyes shut firmly
2. Raise the eyebrows
3. Try to open your eyes very slowly, for a count of approximately 20 seconds
4. When almost fully open, hold for a count of 10
5. Relax

- Repeat

Note:
This is a difficult exercise to master but you will feel those bags lifting.

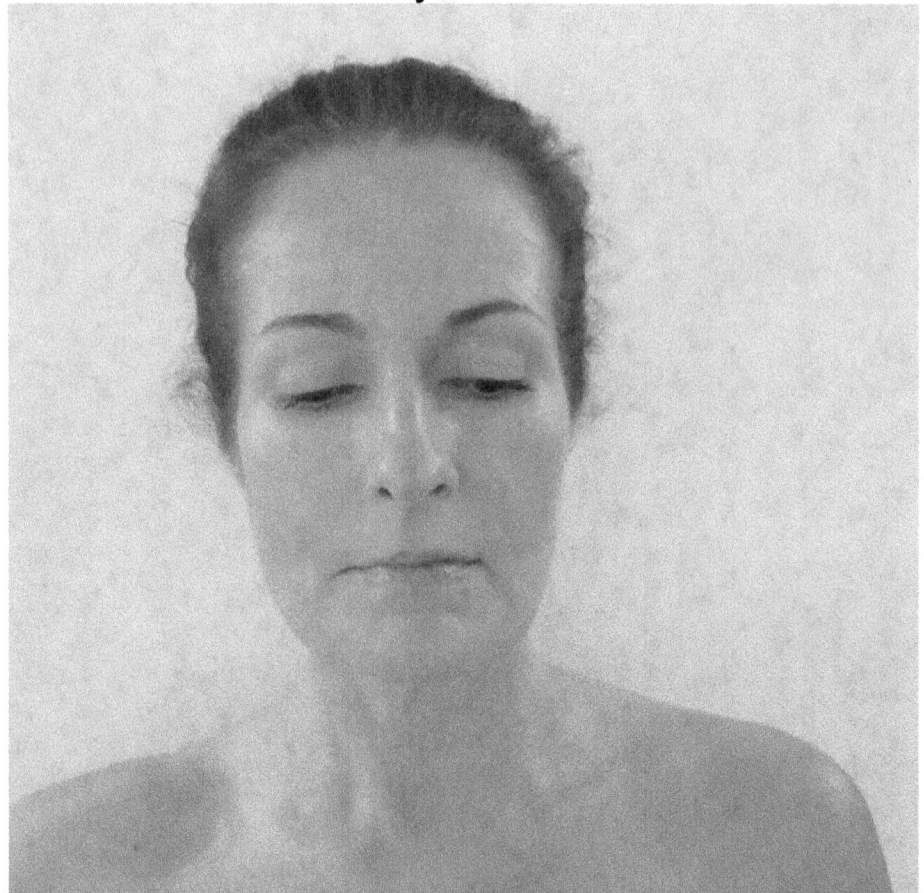

TARGET All around the eyes, both corners, plus under eyelids

1. Lift your eyes up and rotate slowly clockwise for 3 revolutions
2. Repeat anticlockwise

Note:
Each revolution should take about 10 seconds. Extend your glance fully in each direction.

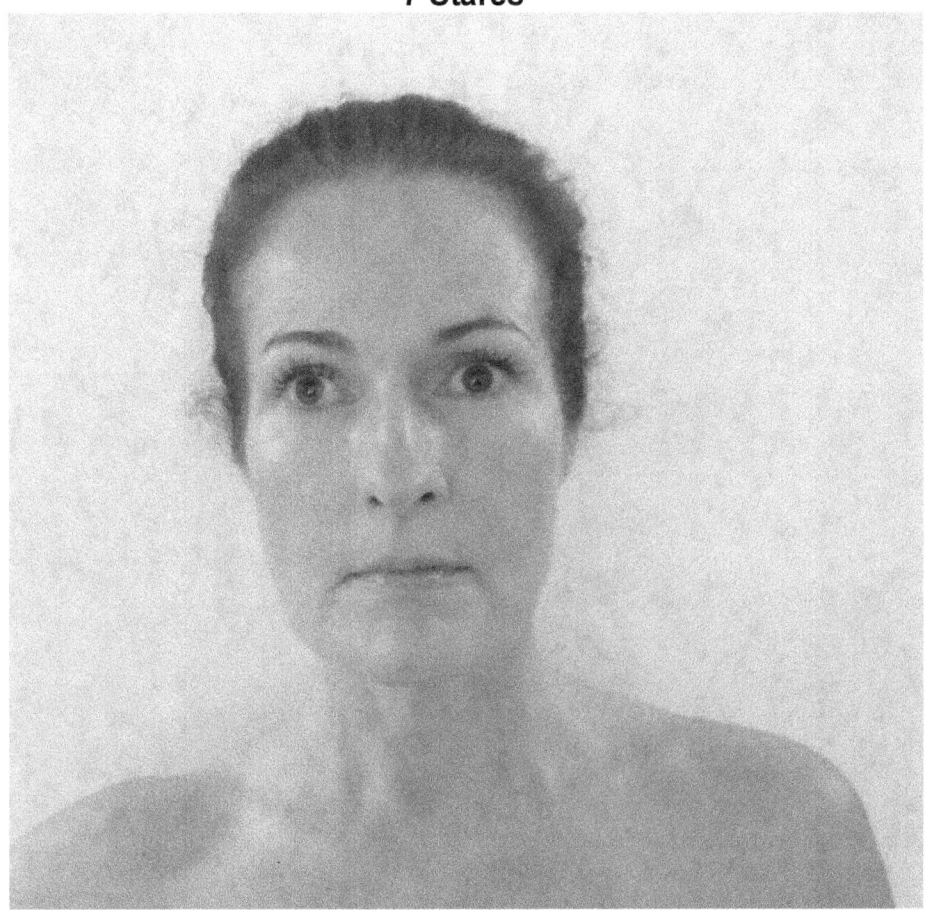

TARGET Eyes, temples & cheeks

1. Open your eyes wide and stare at a distance object
2. Hold as long as you can blink, when you have, relax
3. Start again.
4.
• Continue doing this for approximately1minute

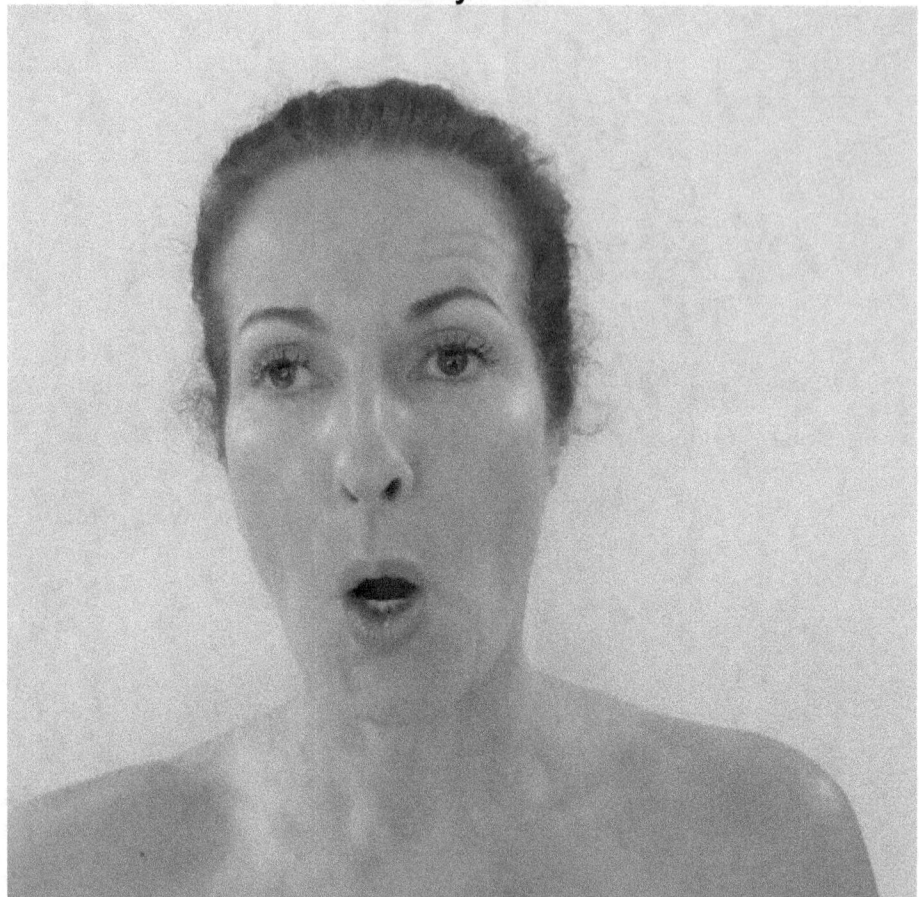

TARGET Sides of nose, nasolabial folds & upper lip

1. Open your mouth slightly and crinkle your nose
2. Slowly make forward circular motions, up forwards, down and around with the tip of your nose and top lip

- Repeat 20 times in one direction and then reverse
- Continue doing this for approximately1minute

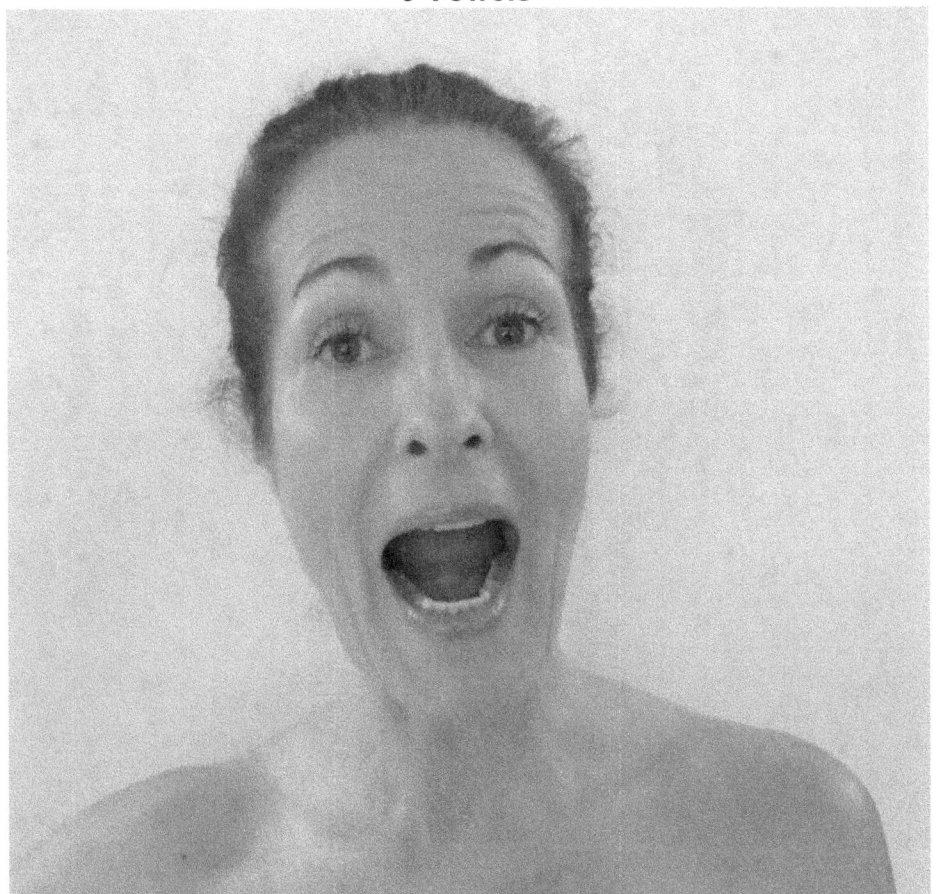

TARGET Lips, cheeks & neck

1. Open your mouth wide and slowly say the vowels A, E, I, O, U in an exaggerated manner

- Repeat 10 times

Note:
You should feel a stretch with each new movement.

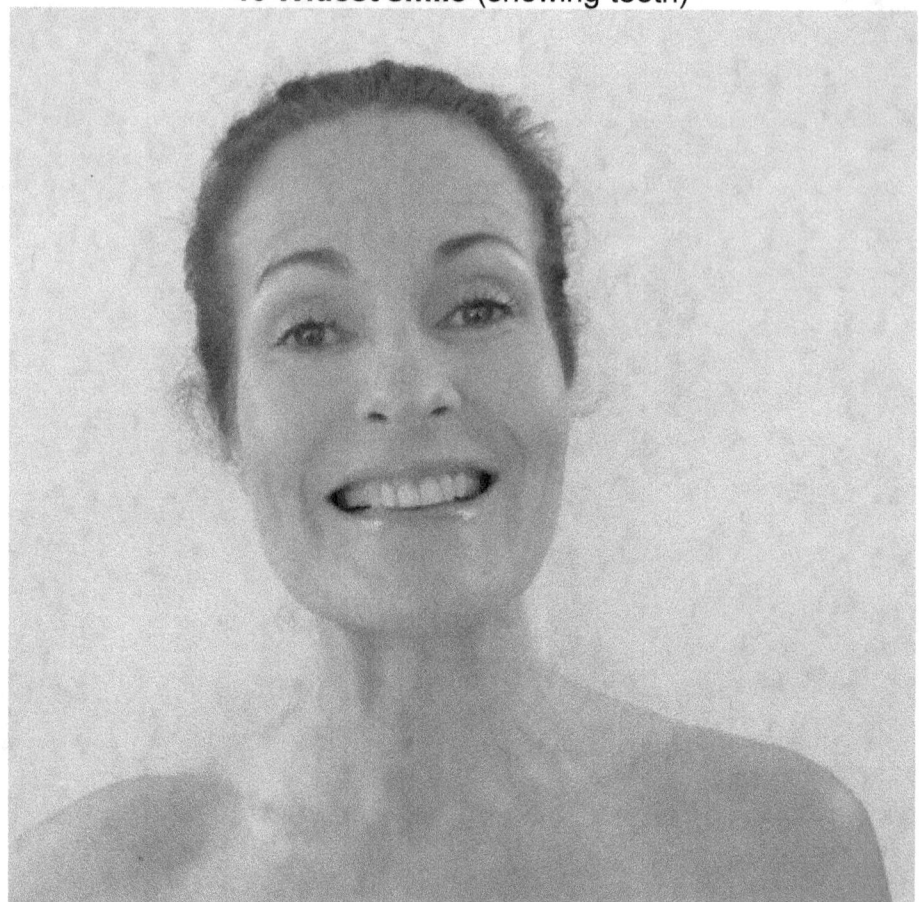

TARGET Cheeks, jowls & chin

1. Make a huge happy broad smile, parting your lips, pulling the lips as wide open to the sides as they will go
2. Lift eyebrows

• Hold wide open for 1 minute

Note:
The smile should extend to the corners of your eyes.

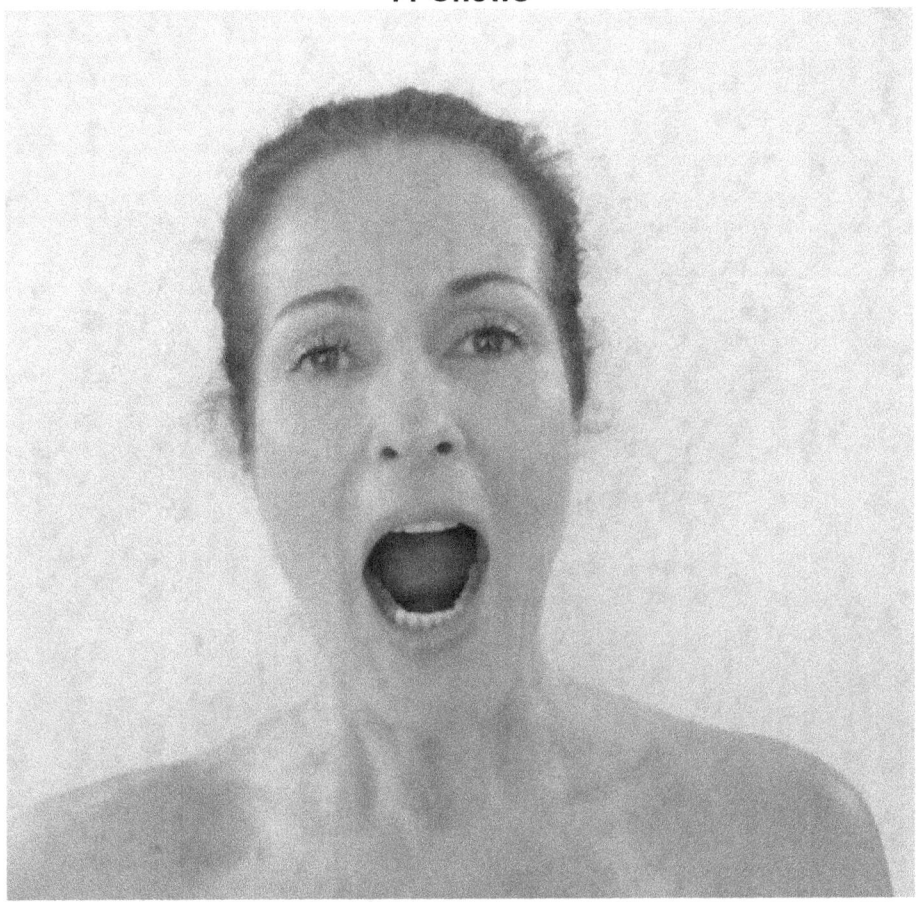

TARGET Cheeks & jowls

1. Open your mouth
2. Smile slightly
3. Chew in reverse slowly with a wide open mouth

- Repeat 50 times

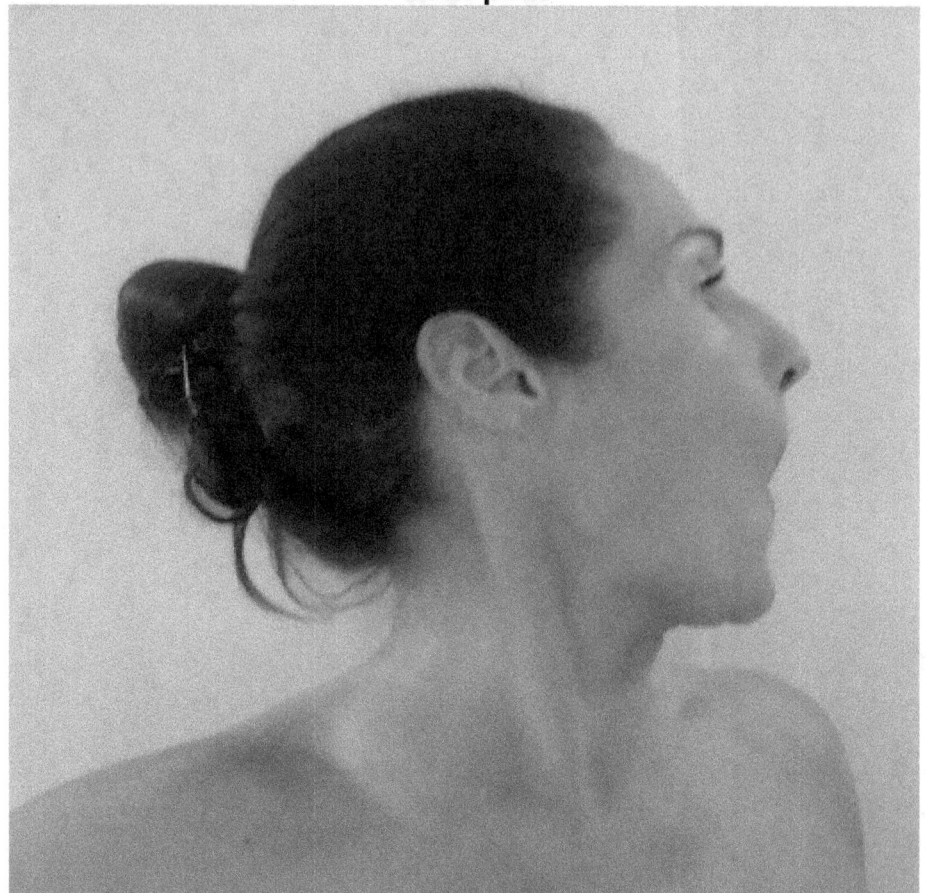

TARGET Jowls & nasolabial folds

1. Turn your head to one side
2. Pull your lips to that side
3. Open and close your mouth slowly but continuously 10 times
4. Relax

- Repeat to the other side

ANYTIME
13 Ceiling kisses

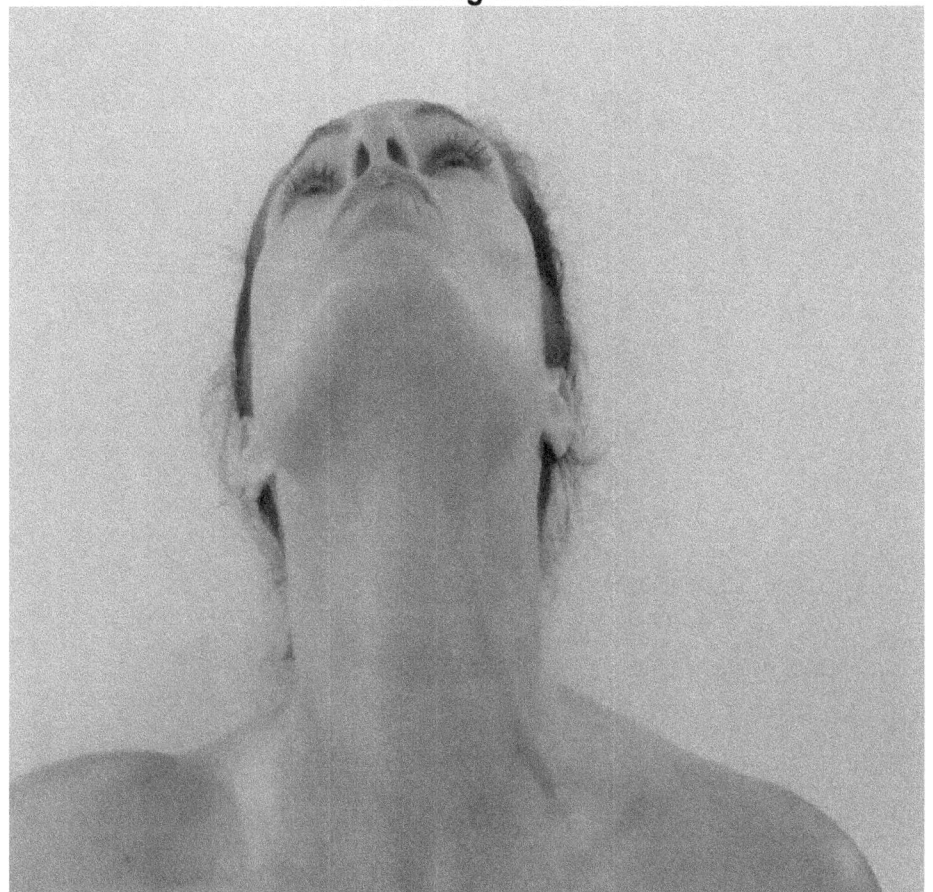

TARGET Neck, chin & sides of mouth

1. Sit upright, feet square, relax your shoulders
2. With your mouth closed, gently extend your neck before tilting your head backwards and let it rest
3. Look at the ceiling
4. Pucker your lips and slowly extend them upwards as if to kiss the ceiling
5. Hold for a count of 10
6. Relax
7. Bring your head slowly forward

• Repeat 3 times

Note:
You will feel this in your neck and possibly your cheeks.

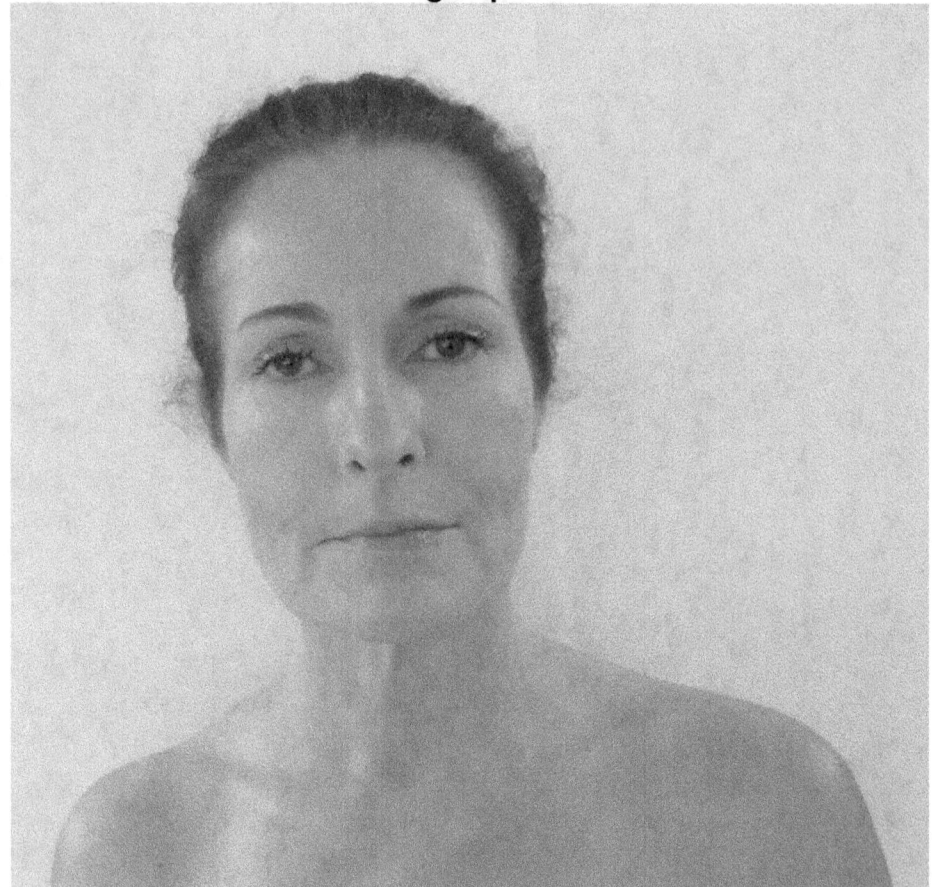

TARGET Under chin

1. With closed lips and teeth gently clench
2. Push your tongue up into the roof of your mouth.
3. Hold for approximately 30 seconds, release your tongue, relax

- Repeat once

Note:
You should feel the muscles under your chin and at the sides of the jaw working.

ANYTIME
15 Scalp relaxations

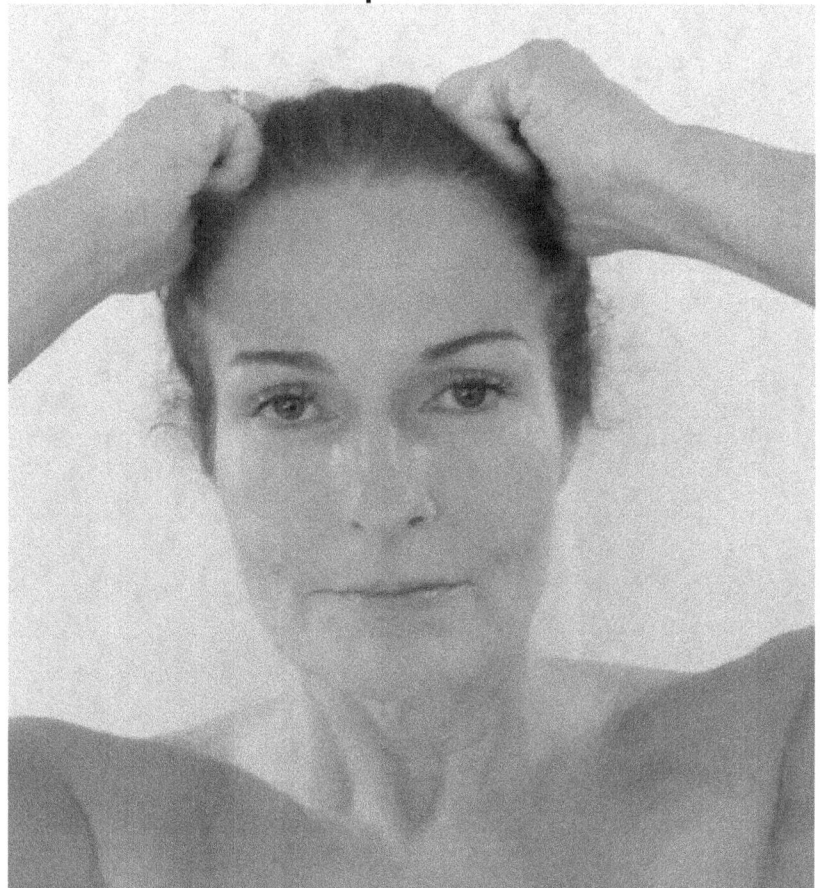

TARGET Scalp & everywhere!

1. Grasp your hair by the roots at the temples
2. Gently, firmly pull the hair making small slow circular rotations
3. Make 3 clockwise rotations and then another three circles anti-clockwise
4. Move your hands a little further away and repeat

- Continue moving your hands over your scalp for about a minute

Chapter 4

Step 2

Standard Fitface toning programme

The **Standard** routine was designed to be the next stage up from the Anytime **Fitface toning** programme and to be the intermediate programme. Therefore this stage should not be attempted until the Anytime programme has been achieved and maintained for more than six weeks. This will allow your muscles to build and become familiar with some of the fundamental toning exercises before introducing these more complex movements.

Do not progress to the next programme (Advanced) until you have given your muscles a chance to become accustomed to the new programme which generally takes a month. The routine should be used in conjunction with the other programmes to give your face a balanced training routine. The idea of the 3 step **Fitface toning** programme is to give your face toned appearance, not a body builder's taut face.

There are three levels to progress through: Beginners, Regular and Target. The first two levels in each programme must be completed before continuing to the next stage. The Target level of each programme is optional; it is the Regular level but more intense with the opportunity to select additional exercises to target your own personal problem areas.

Beginner's level

Week one
Either cut the repetitions or the duration of each exercise in half.
Only when you are ready, continue to the next **Regular level.**

Regular level

Continue with the full programme for the next six weeks.
Only when you are ready, continue to the last target level of the programme or change to the Standard **Fitface toning** programme.

Target level

Select any 3 additional toning exercises from:
The Anytime, Standard or Advanced **Fitface toning** programmes, which you feel would be most beneficial for you, to target your personal problem areas.

Notes:

Timings: are for guidance only
Count: means timing in approximate seconds

Duration

This depends totally on how familiar you are with the exercises. In round terms I think of each exercise as taking approximately 1 minute. Each programme has 14/15 exercises with the **Micro** or **Face-lift** taking 3 minutes. Therefore I would expect you to spend approximately 20 minutes on each programme for maximum results.

STANDARD
1 Big O kiss

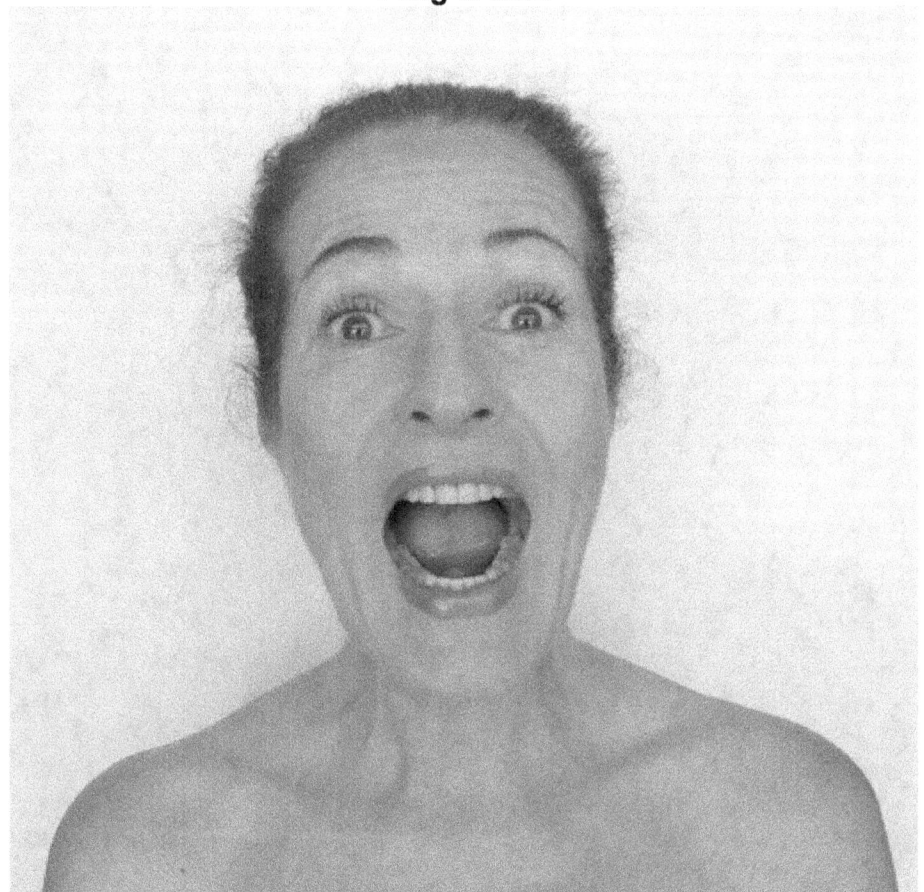

TARGET Warm up - whole face

1. Open your mouth as wide as you can
2. Raise your eyebrows (pulling everything back)
3. Pause for a moment
4. Slowly bring your lips in to form a kiss with your eyebrows into a slight frown
5. Pause for a moment

• Repeat continuously for 20 repetitions

STANDARD
2 Eyebrow lifts (closed eyes)

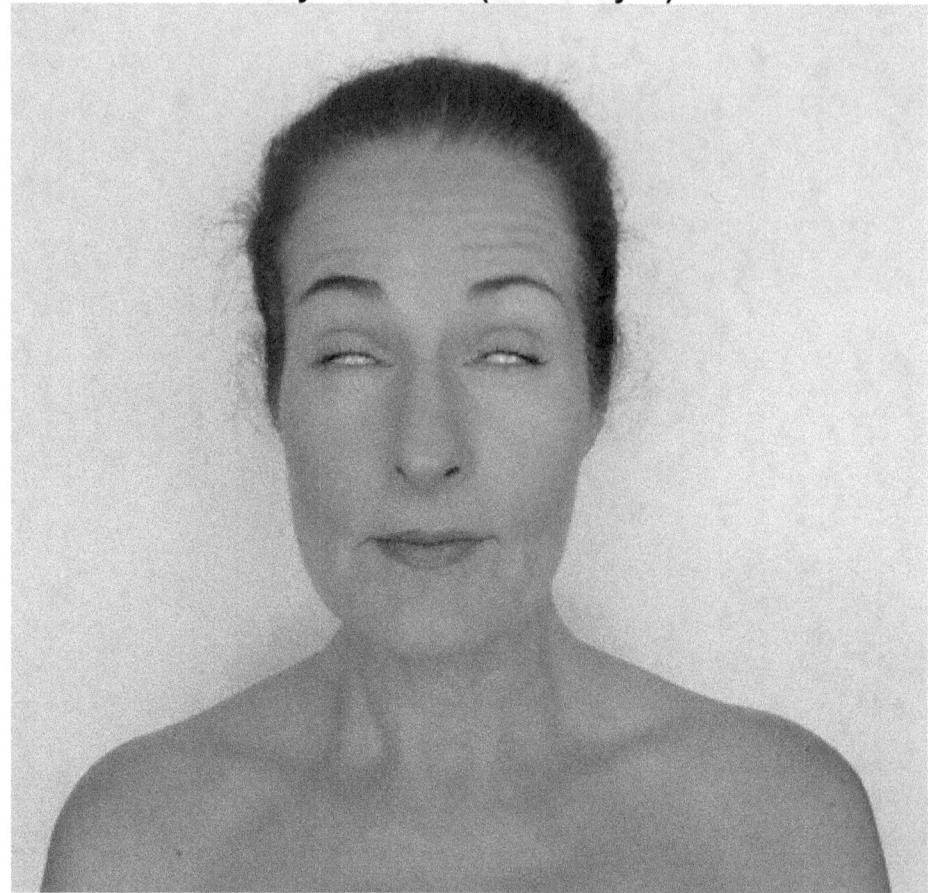

TARGET Brow, eyebrows & upper eyelid

1. Close your eyes
2. Raise your eyebrows as high as possible and then lower slowly but continuously

- Repeat 30 times
- Total 1 minute

Note:
Resist the urge to open your eyes. Should they open momentarily as mine have, just close them again.

STANDARD
3 Face-lift

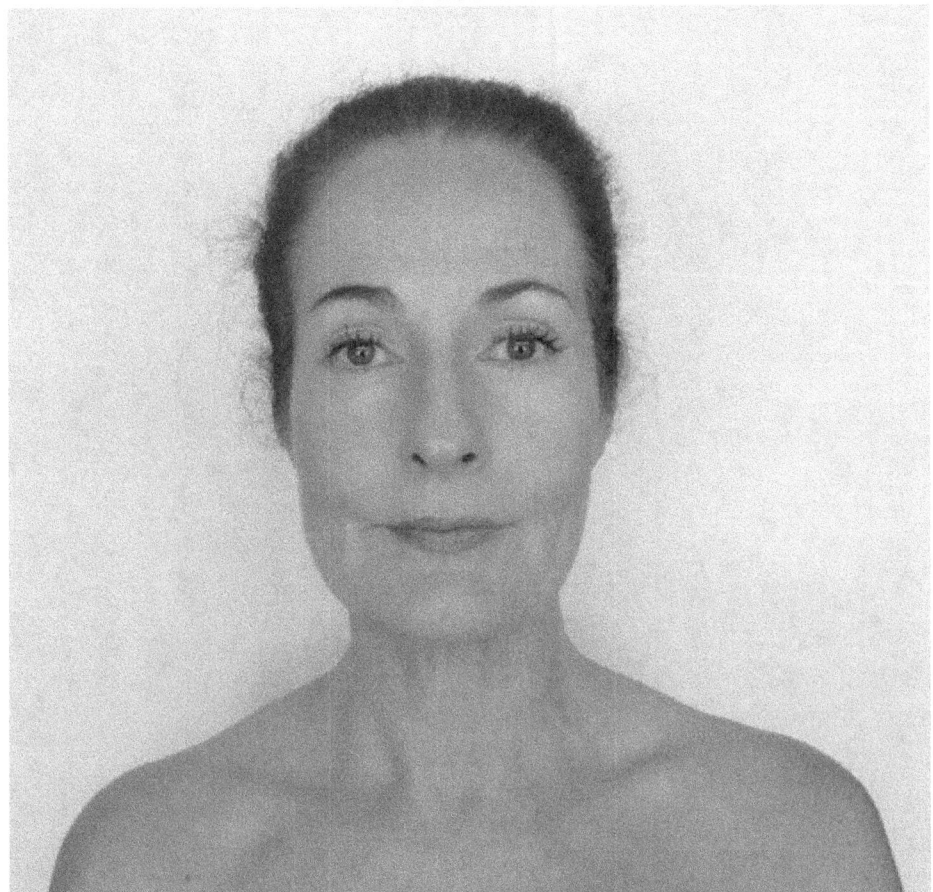

TARGET Whole face, especially forehead and cheekbones

1. With eyes open (easier) or shut
2. Lift eyebrows
3. Contract muscle of scalp and temples, lifting you eyebrows up and back to connect with the back of the neck

- Hold for I minute
- Repeat twice (i.e. for a total of 3 times)

Note:
Feel the whole face pull taut, up and back. Imagine that your whole skull is being tightened, lifted and pulled back.
Tip: Ensure that the sides of your mouth are level with the corners slightly turning up. **DO NOT FROWN**

STANDARD
4 Stares

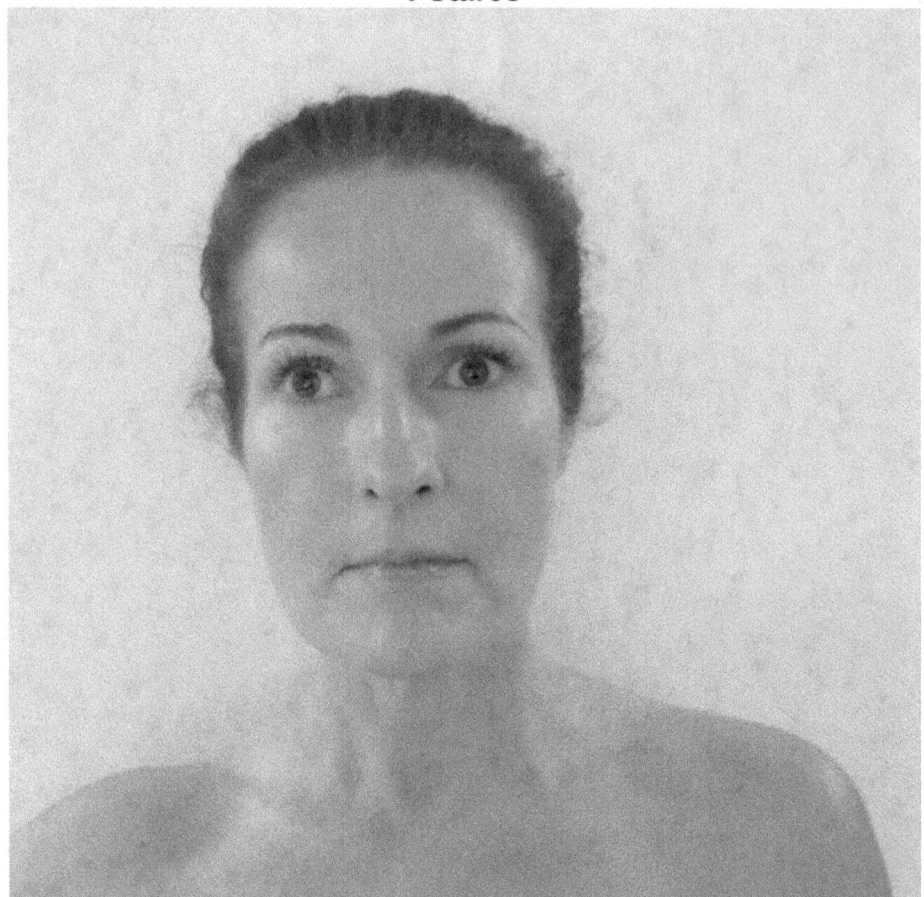

TARGET Eyes, temples & cheeks

1. Open your eyes wide and stare at a distance object.
2. Hold as long as you can. Blink, when you have to and start again

- Continue doing this for approximately 1 minute

STANDARD
5 Eye pulls

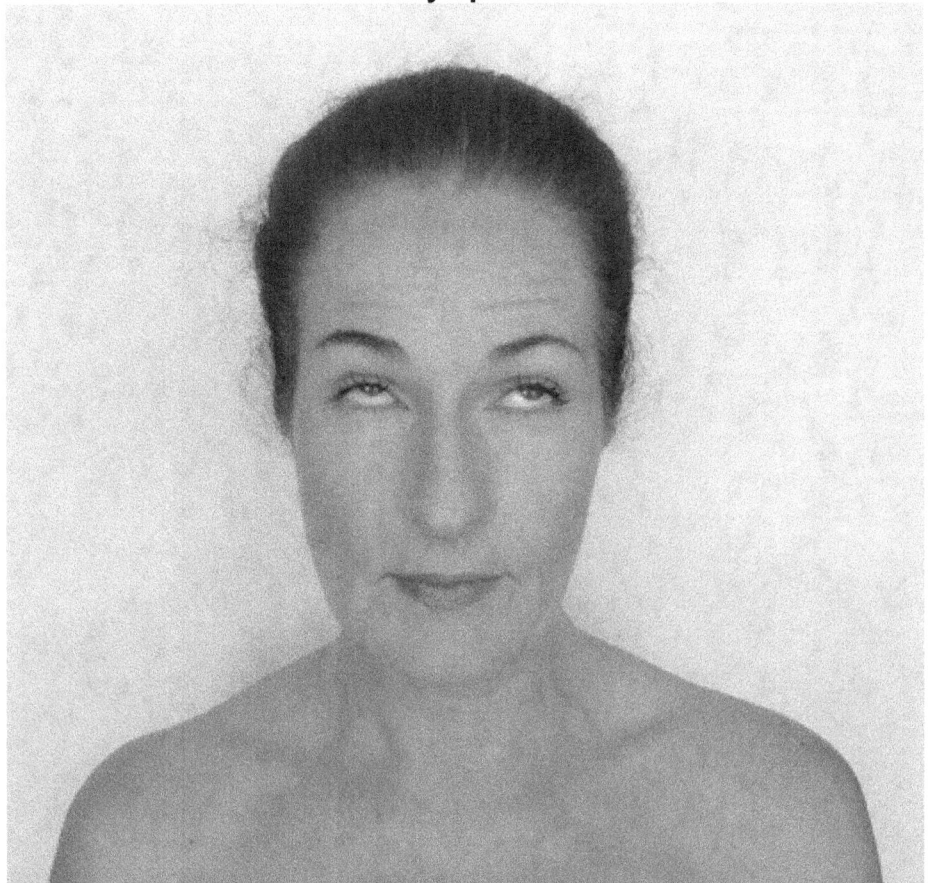

TARGET All around the eyes: both corners, plus under eyelids

1. Begin with eyes shut firmly
2. Raise the eyebrows
3. Try to open your eyes very slowly, for a count of approximately 20 seconds
4. When almost fully open, hold for a count of 10
5. Relax

- Repeat

Note:
This is a difficult exercise to master but you will feel those bags lifting.

STANDARD
6 Eye clocks

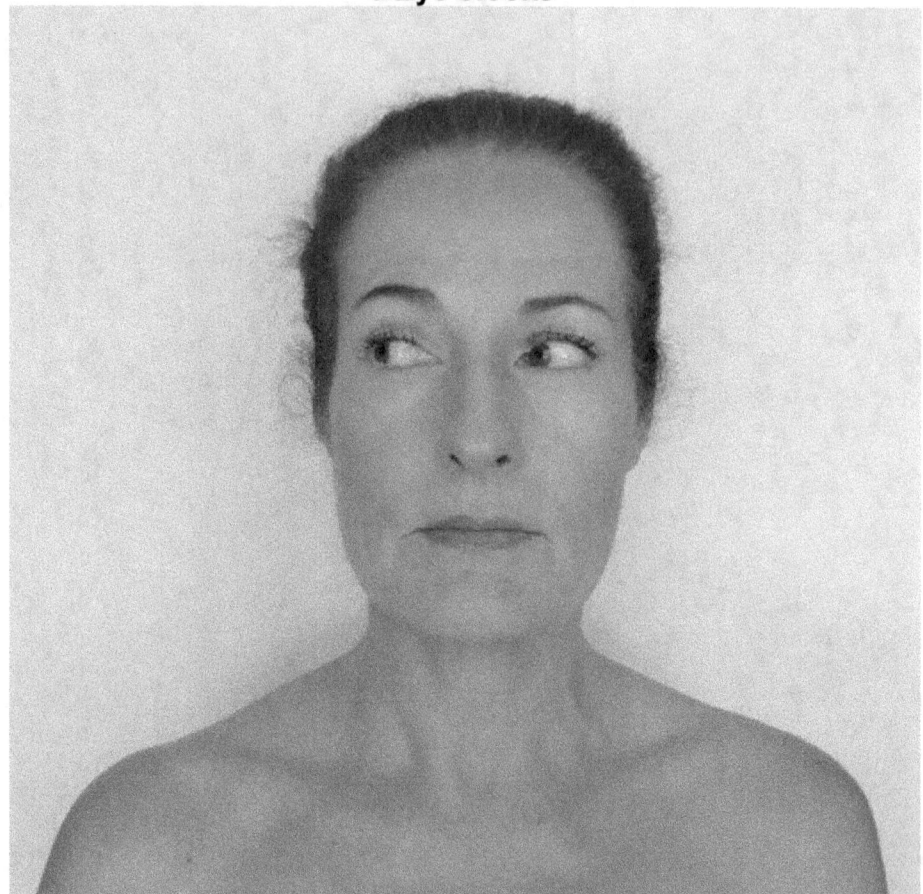

TARGET All around the eyes:
inner & outer corners, plus under eyelids

1. Lift your eyes up and rotate slowly clockwise for 3 revolutions
2. Repeat anticlockwise

Note:
Each revolution should take about 10 seconds. Extend your glance fully in each direction.

7 Bunny noses

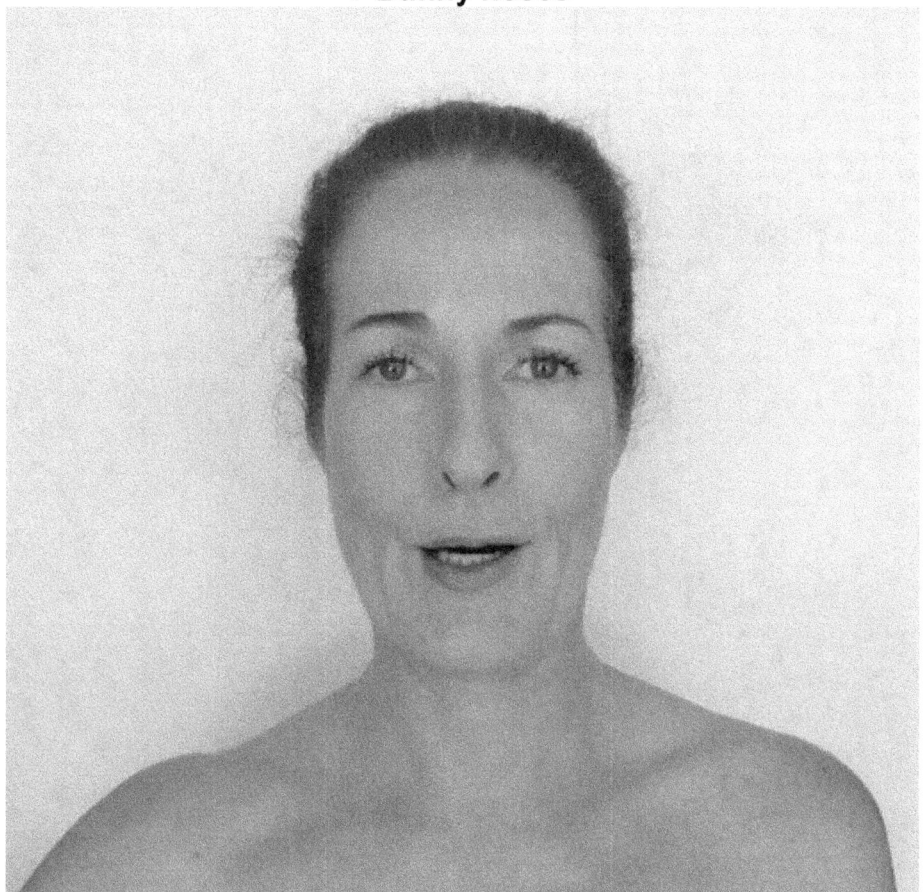

TARGET Sides of nose, nasolabial folds & upper lip

1. Open your mouth slightly and crinkle your nose
2. Slowly make forward circular motions, up forwards, down and around with the tip of your nose and top lip

- Repeat 20 times in one direction and then reverse
  Continue doing this for approximately1minute

STANDARD
8 Three smiles

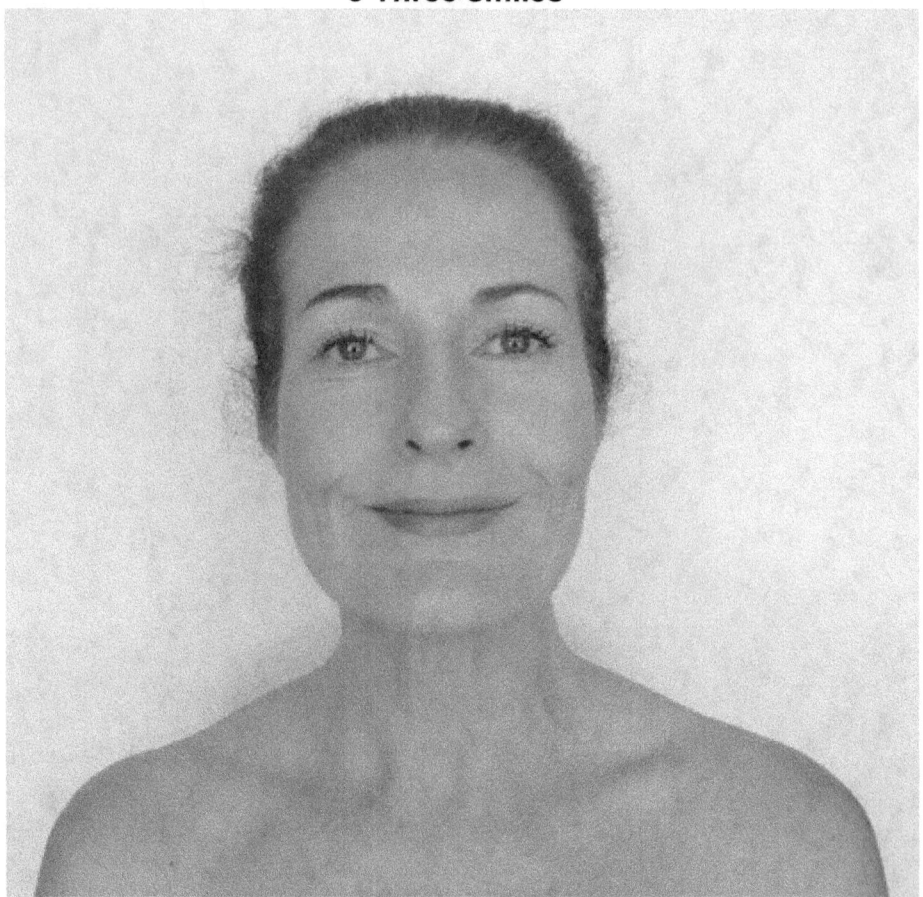

TARGET Cheeks & jowls

1. Lips together smile horizontally, hold count to 10
2. Lips together smile more towards the earlobes, hold count to 10
3. Lips together make a huge smile towards the temples, hold count to 10
4. Still holding taut, reverse each move - i.e. smile towards the earlobes and hold for a count of 10

- Repeat
  i.e. smile horizontally and hold for a count of 10
  i.e. smile towards the earlobes and hold for a count of 10

STANDARD
9 Widest smile (closed lips)

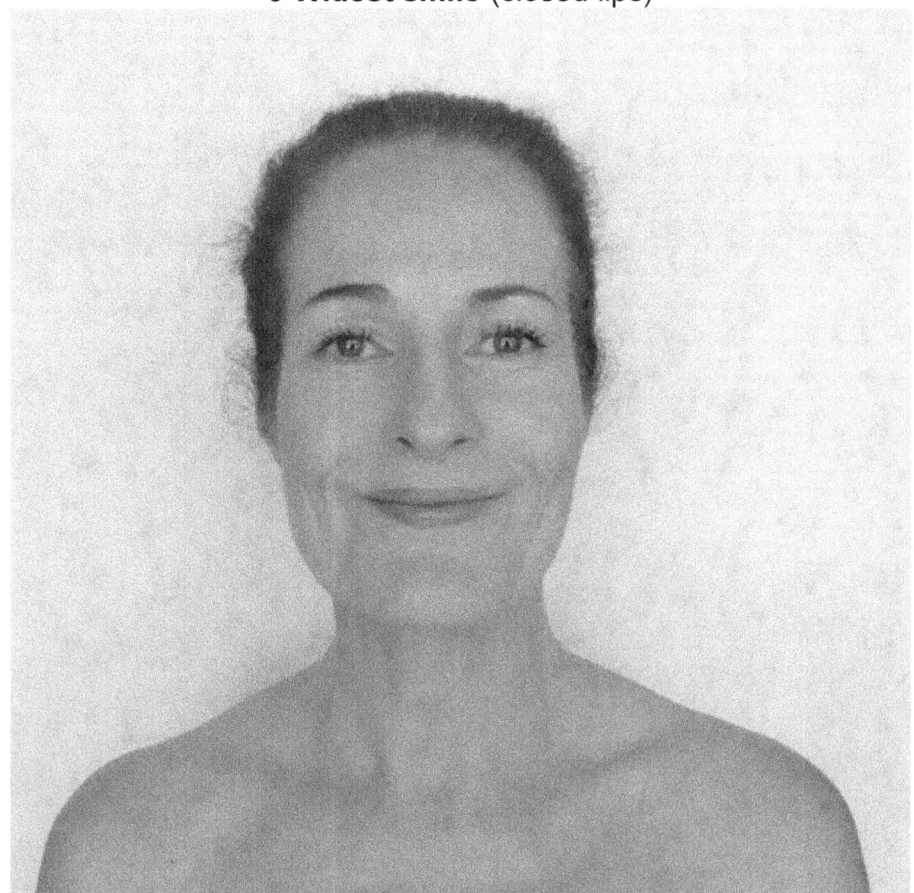

TARGET Cheeks, jowls & chin

1. Make a huge happy broad smile, with closed lips, pulling the lips as wide open to the sides as they will go.
2. Lift eyebrows

- Hold wide open for I minute

Note:
The smile should extend to the corners of your eyes.

STANDARD
10 Nose pulls

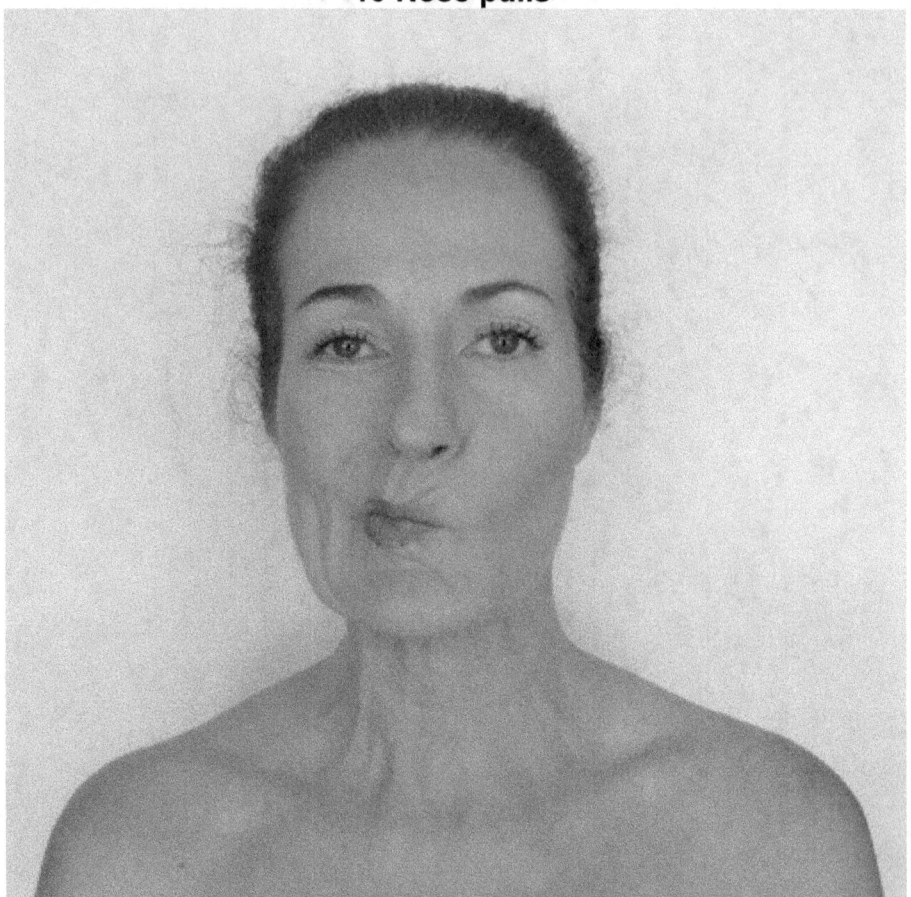

TARGET Lips and cheeks

1. Pucker your lips into a pursed kiss
2. Very slowly pull them as far to one side of the face as you can comfortably, pause, but do not release the tension
3. Still holding your lips taut move them to the other side of the face, the corner of the mouth should be turned upwards

- Repeat 6 times

Note:
This is difficult, you may have to stop half way and release the tension and start again.

STANDARD
11 Tongue out

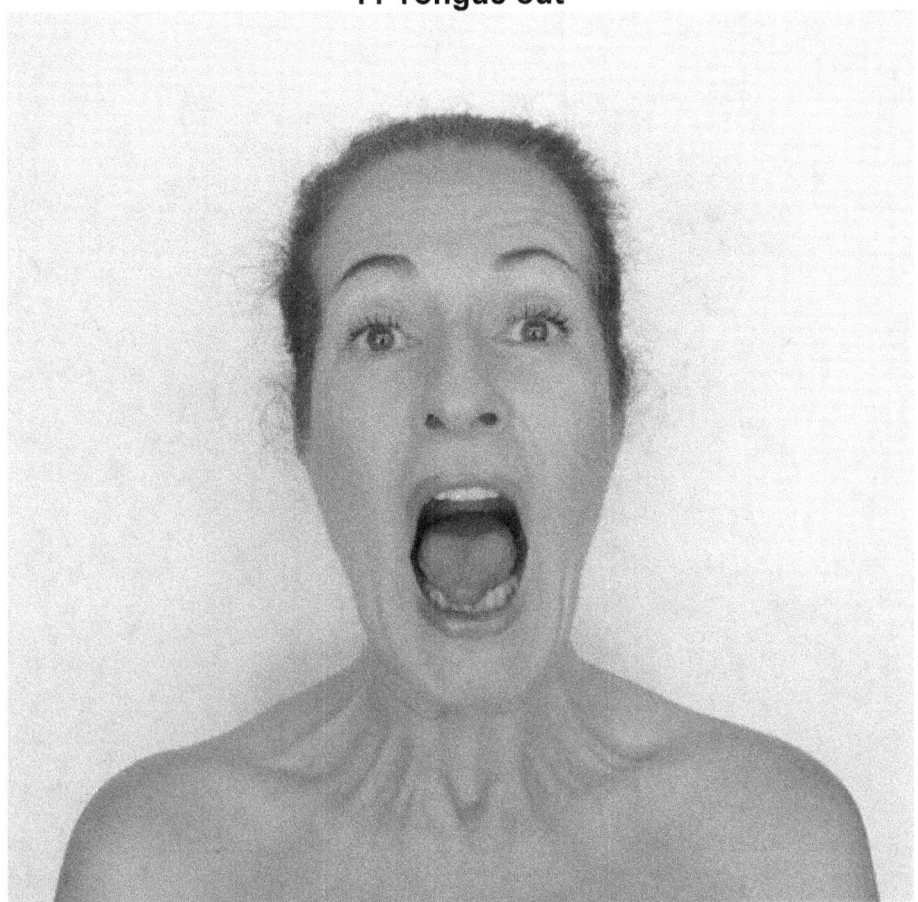

TARGET Lips, neck & chin

1. Open your mouth as wide as possible
2. Stick your tongue out as far as possible
3. Raise you eyebrows
- Hold for a count of 25

- Repeat

Note:
There may be a tendency to pull the lips forward after a while, try to resist by pulling them up as if smiling at the corners of the mouth. This is harder than it looks, be patient with yourself, only hold for a count of 15 at first, don't over strain.

STANDARD
12 Jowl pull smile

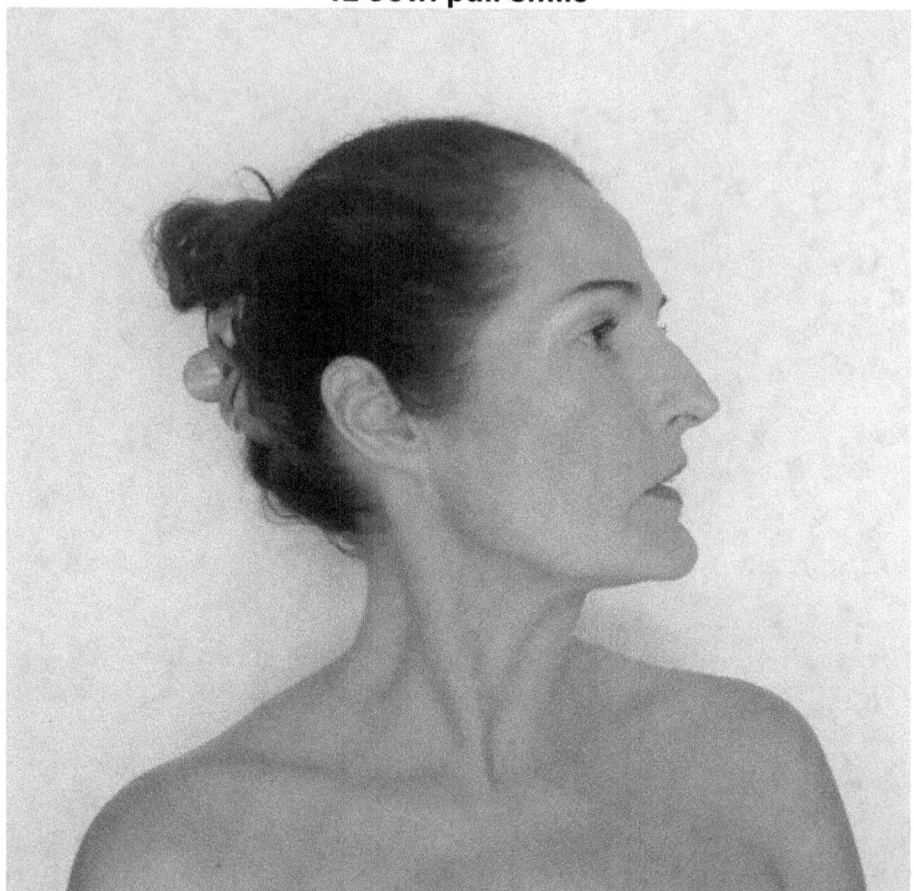

TARGET Jowls & nasolabial folds

1. Turn your head to one side
2. Open your mouth slightly
3. Pull your lips to that side; they should automatically pull up to the temple, pause and hold taut for a count of 30
4. Relax

- Repeat to the other side

STANDARD
13 Jaw circles

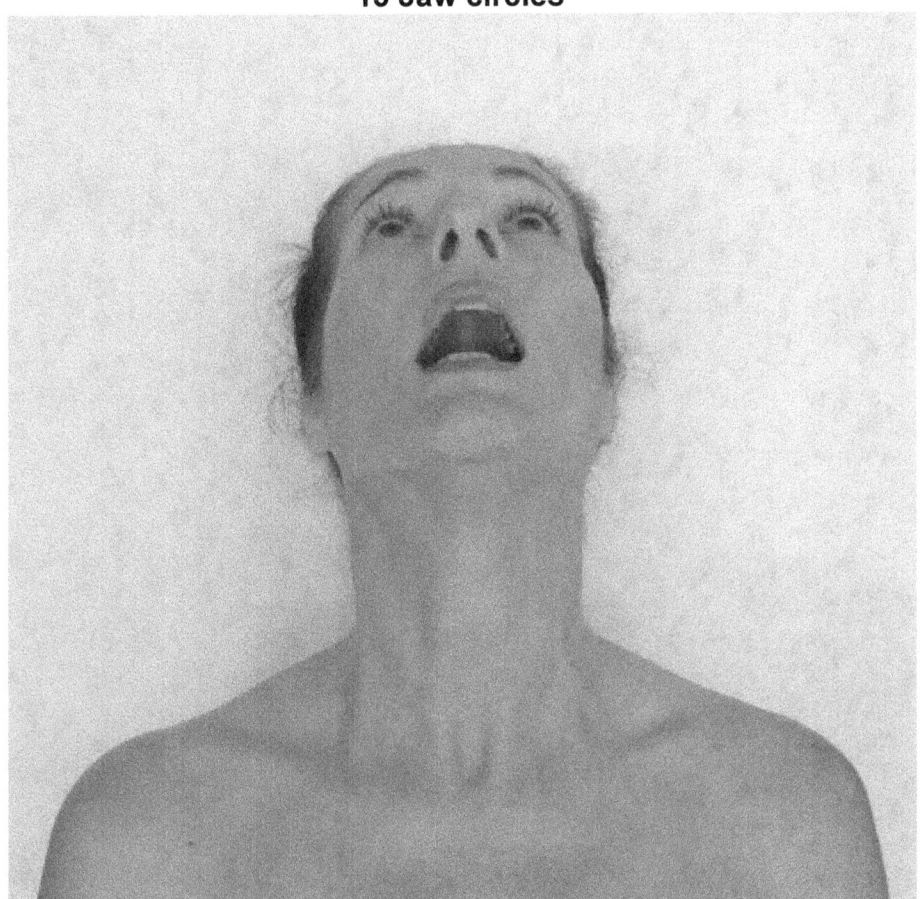

TARGET Chin & neck

1. Sit upright, relax your shoulders with feet firmly planted on the floor
2. Extend your neck and gently tilt your head backwards looking at the ceiling in front of you
3. Open your mouth wide and start a slow chewing movement forwards 10 times
4. Pause and reverse movement, chewing backwards slowly 10 times
5. Slowly bring your head to an upright position

- Repeat

STANDARD
14 Chin slaps

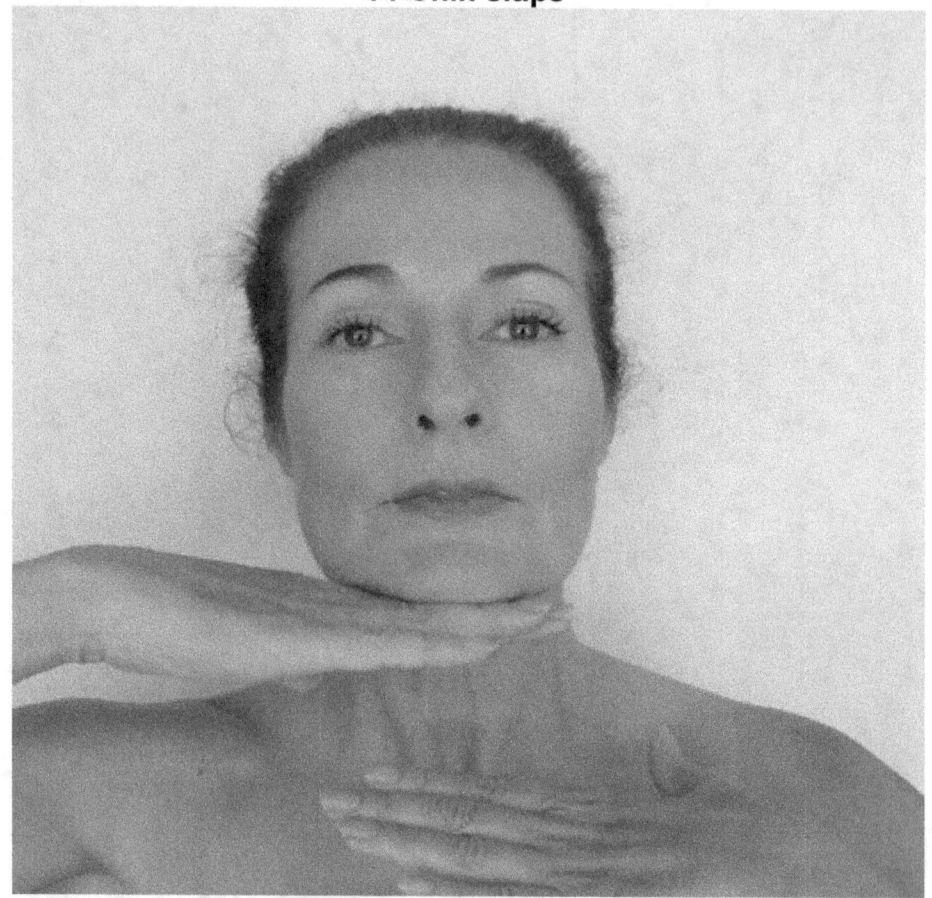

TARGET Jowls & cheeks

1. With the backs of your hands, firmly slap the underneath of your chin, following your jaw line from one side to the other side and back again

- Repeat continuously for about a minute

STANDARD
15 Ears relaxation

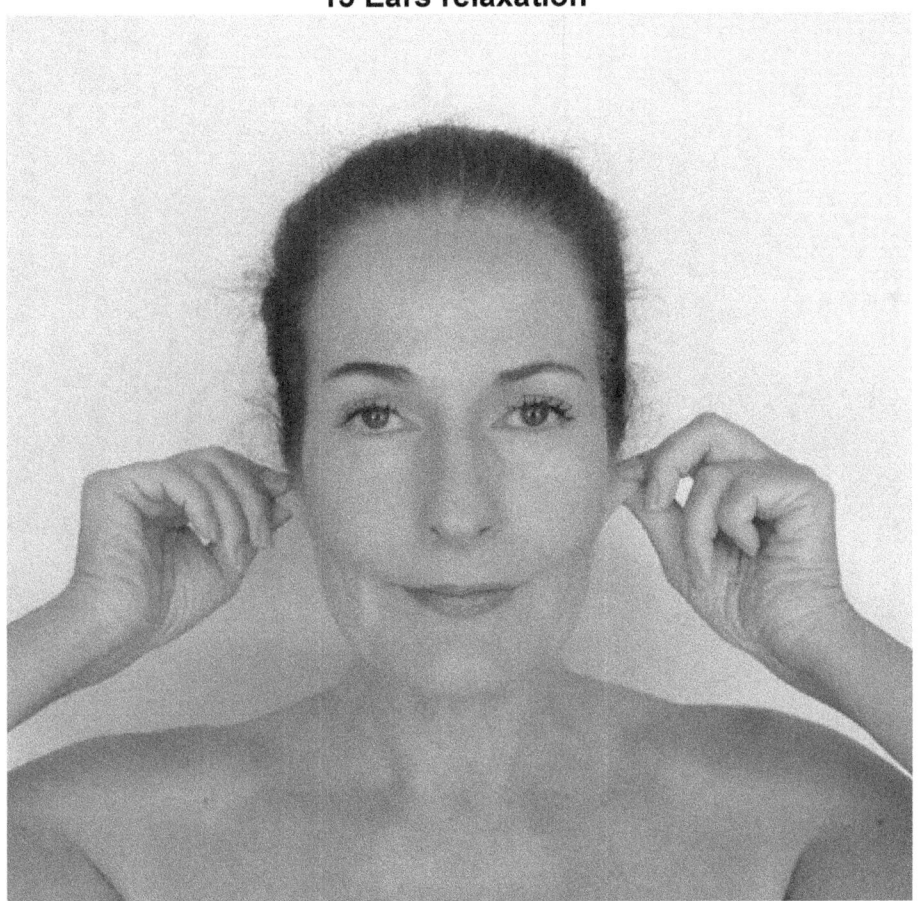

TARGET Ears & jowls

1. Pinch your earlobes and massage between fingers moving up the back of the ear coming all the way to the front
2. Put your index finger in your ear hole and gently rotate your finger to massage inner ear whilst your thumb gently massages the back of the outer ear

Chapter 5

Step 3

Advanced Fitface toning programme

The Advanced routine was designed to be the next stage up from the Standard **Fitface toning** programme and the last stage. Therefore this stage should not be attempted until the Standard programme has been achieved and maintained for at least six weeks. This will allow your muscles to build and become familiar with some of the central toning exercises before introducing these more complicated movements.

This routine should be used in conjunction with all the other programmes to give your face a balanced training routine. The idea of the 3 step **Fitface** is to give your face a toned appearance.

There are three levels to progress through: Beginners, Regular and Target. The first two levels in each programme must be completed before continuing to the next stage. The Target level of each programme is optional; it is the Regular level but more intense with the opportunity to select additional exercises to target your own personal problem areas.

Beginner's level

Week one
Either cut the repetitions or the duration in half for each exercise.
Only when you are ready, continue to the next **Regular level**.

Regular level

Continue with the full programme for the next six weeks.
Only when you are ready, continue to the last target level.

Target level

Select any 3 additional toning exercises from:
The Anytime, Standard or Advanced **Fitface toning** programmes

which you feel would be the most beneficial for you, to target your personal problem areas.

Notes:

Timings:	are for guidance only
Count:	means approximate seconds

Duration

This depends totally on how familiar you are with the exercises. In round terms I think of each exercise as taking approximately 1 minute. Each programme has 14/15 exercises with the **Micro** or **Face-lift** taking 3 minutes. Therefore I would expect you to spend approximately 20 minutes on each programme for maximum results.

ADVANCED
1 Big O kiss

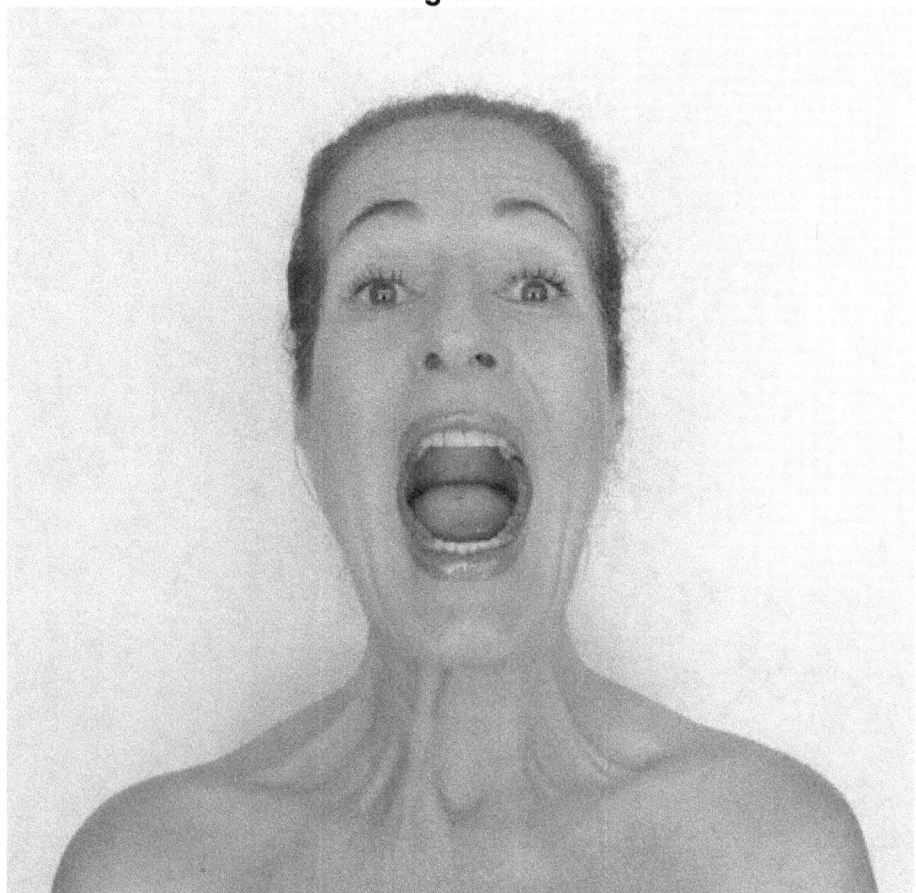

TARGET Warm up - whole face

1. Open your mouth as wide as you can
2. Raise your eyebrows (pulling everything back)
3. Pause for a moment
4. Slowly bring your lips in to form a kiss with your eyebrows into a slight frown
5. Pause for a moment

• Repeat continuously for 20 repetitions

ADVANCED
2 Eyebrow lifts

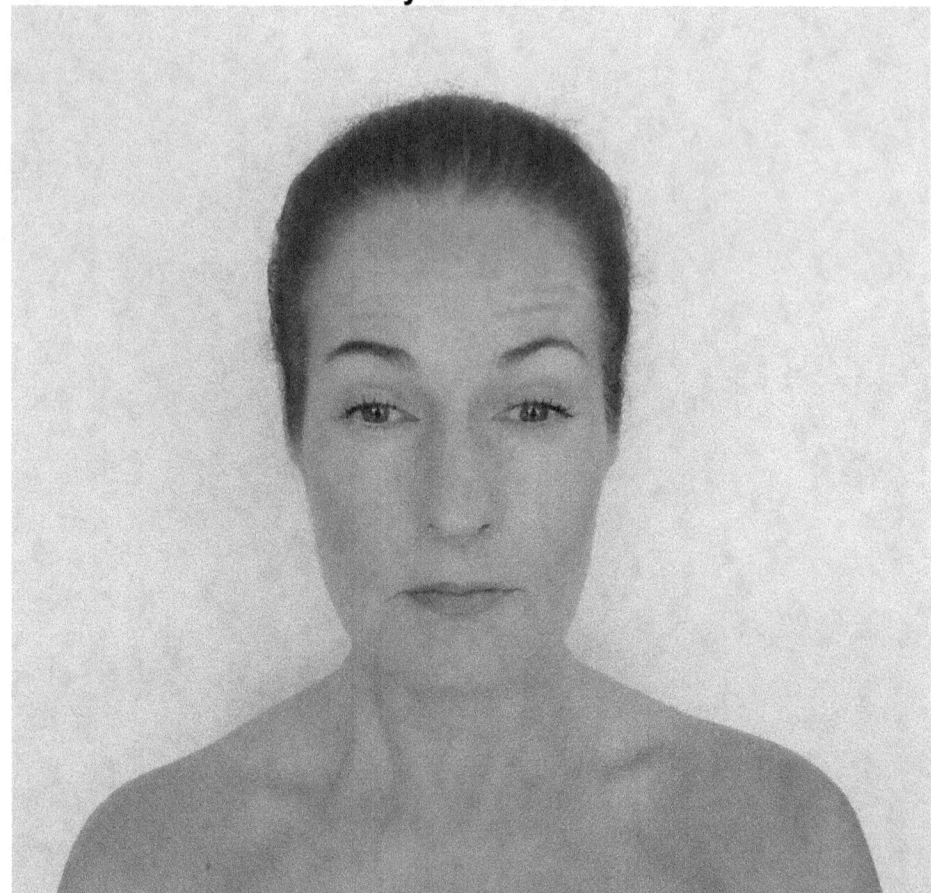

TARGET Brow, eyebrows & upper eyelid

1. Smile gently and hold
2. Lift your eyebrows as high as possible and pause for a couple of seconds
3. Relax

- Repeat approximately 30 times
- Total 1 minute

Note:
Feel as if, you are raising and lowering your eyebrows slowly.

ADVANCED
3 Face-lift

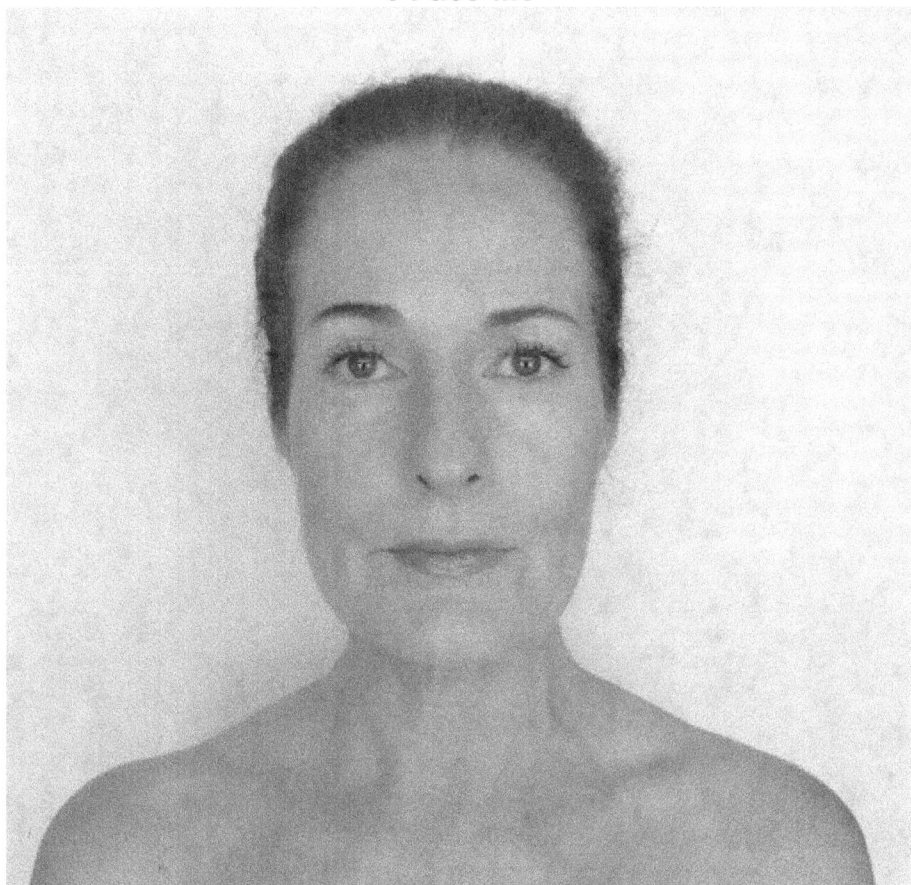

TARGET Whole face, especially forehead and cheekbones

1. With eyes open (easier) or shut
2. Lift eyebrows
3. Contract muscle of scalp and temples, lifting you eyebrows up and back to connect with the back of the neck.
- Hold for I minute
- Repeat twice (i.e. for a total of 3 times)

Note:
Feel the whole face pull taut, up and back. Imagine that your whole skull is being tightened, lifted and pulled back.
Tip: Ensure that the sides of your mouth are level with the corners slightly turning up. **DO NOT FROWN**

ADVANCED
4 Eye pulls

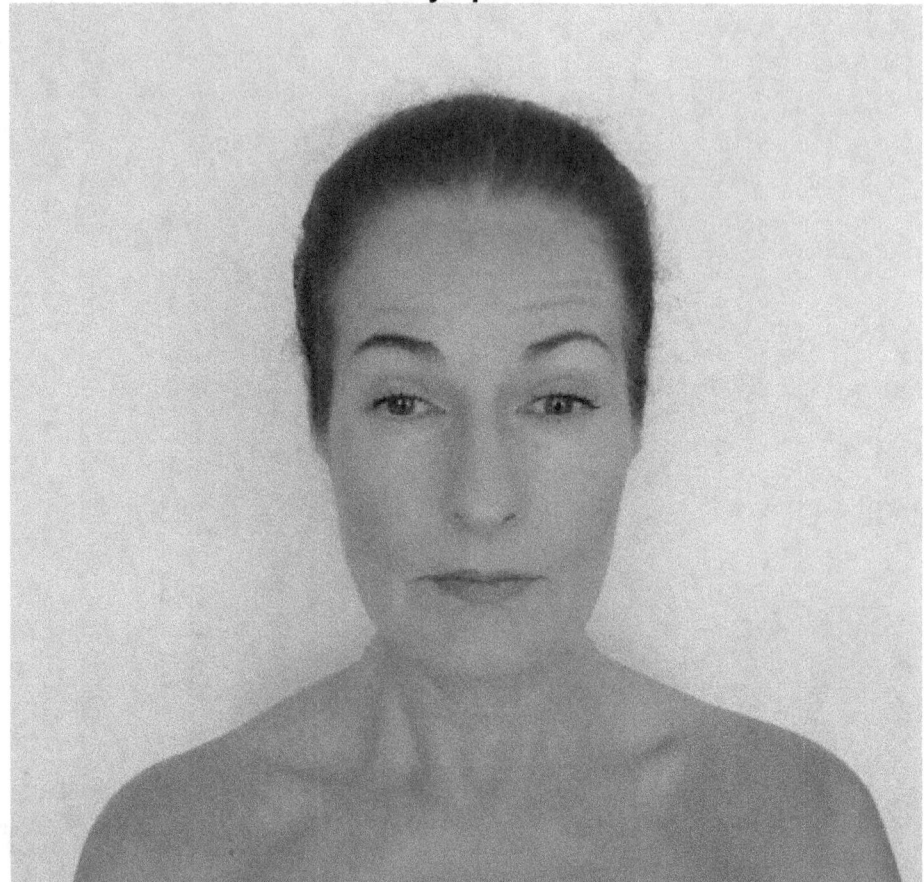

TARGET All around the eyes, both corners, plus under eyelids

1. Begin with eyes shut firmly
2. Raise the eyebrows
3. Try to open your eyes very slowly, for a count of approximately 20 seconds
4. When almost fully open, hold for a count of 10
5. Relax

- Repeat

Note:
This is a difficult exercise to master but you will feel those bags lifting.

ADVANCED
5 Clocks (closed)

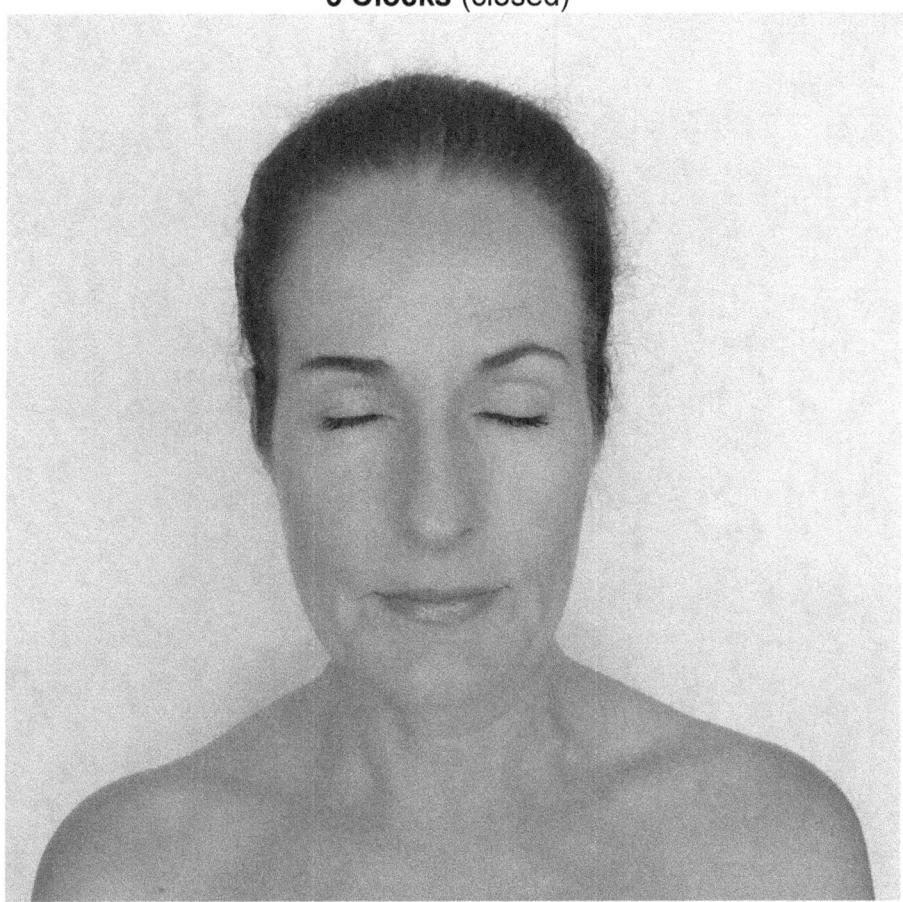

TARGET Sides of eyes

1. Lift your eyes up and rotate slowly clockwise for 3 revolutions
2. Repeat anticlockwise

Note:
Each revolution should take about 10 seconds. Extend your glance fully in each direction.

ADVANCED
6 Bag lifts

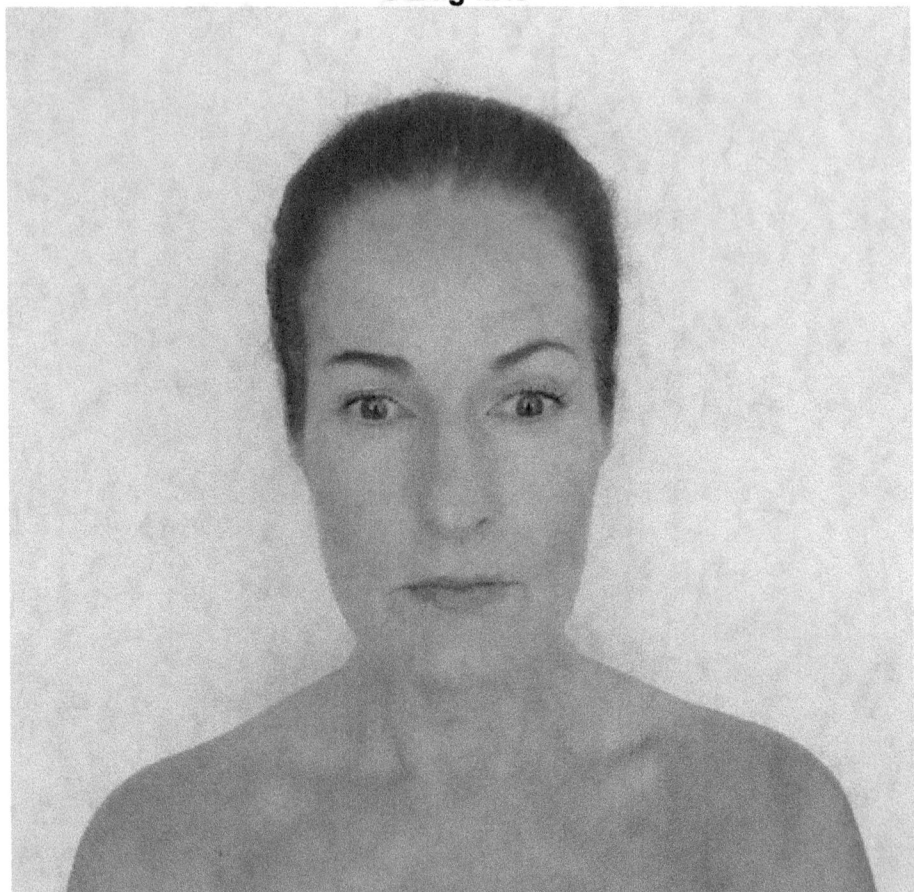

TARGET Sides of eyes

1. Raise eyebrows slightly
2. With eyes open, lift the under eye area as high as possible, pause hold for a count of 5
3. Lower under eyelid
4. Raise and repeat lifting up pause, down

- Repeat 10 times

ADVANCED
7 Bunny noses

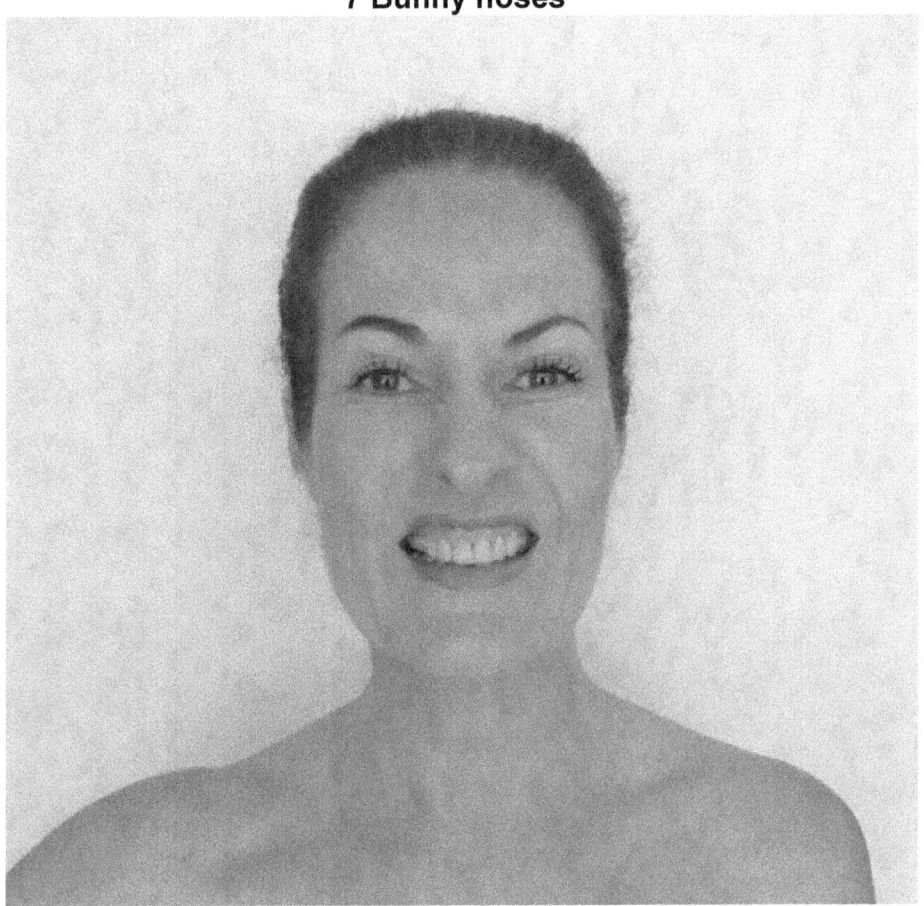

TARGET Sides of nose

1. Open your mouth slightly and crinkle your nose
2. Slowly make forward circular motions, up forwards, down and around with the tip of your nose and top lip

- Repeat 20 times in one direction and then reverse
  Continue doing this for approximately 1 minute

ADVANCED
8 In and out kisses

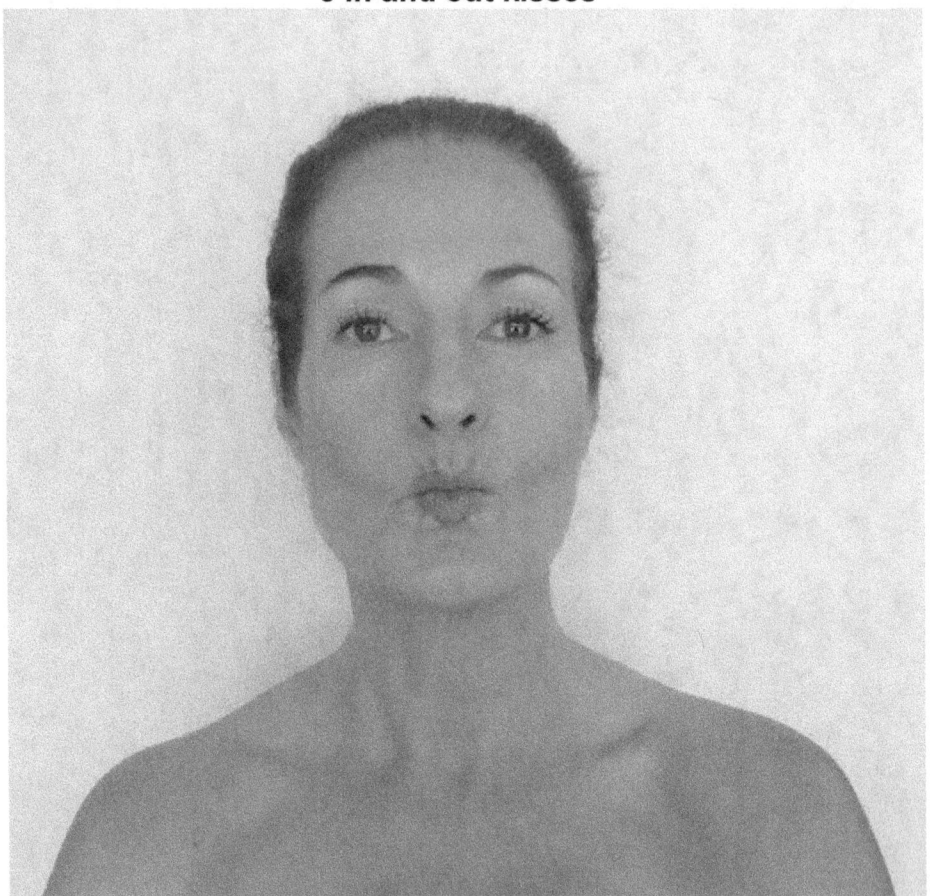

TARGET Sides of nose and cheeks

1. With closed lips, slowly smile broadly to the earlobes
2. Pause for a couple of second
3. Slowly move lips forward into a kiss
4. Pause, for a couple of seconds, keep the tension repeat continuously

- Lift eyebrows (if you want an extra pull)
- Repeat continuously for about 1 minute

Note:
Try to ensure that the corners of the mouth are upright.

ADVANCED
9 Umm

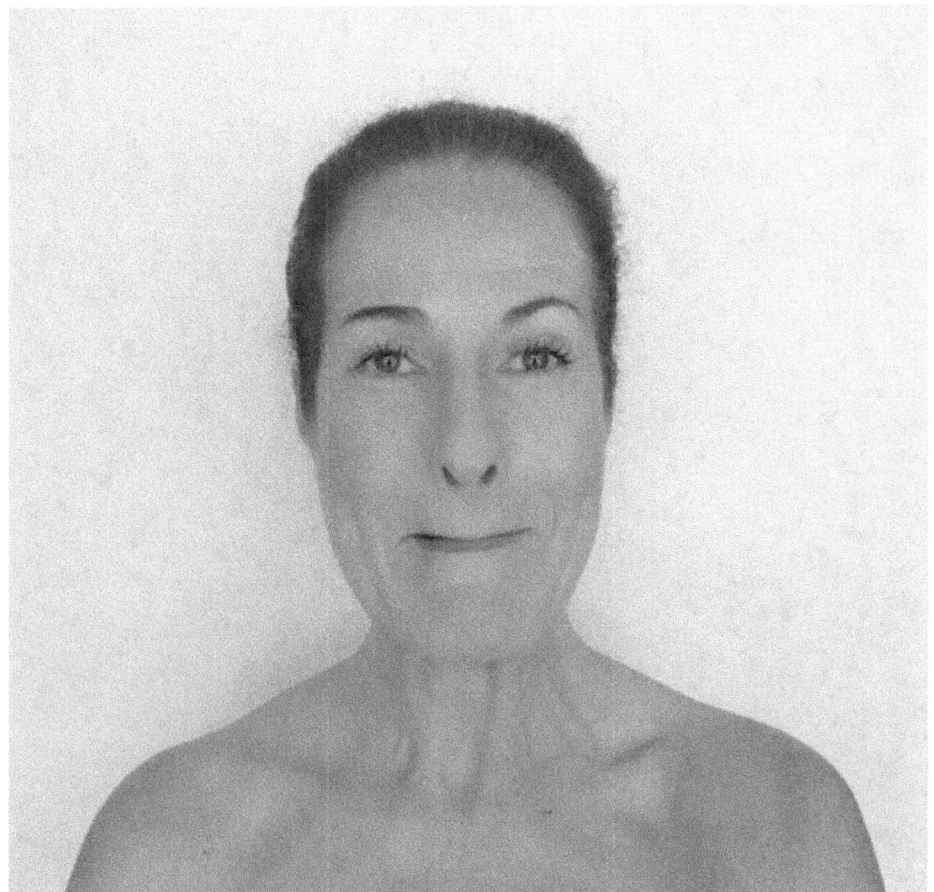

TARGET Cheeks, under eye area & eyebrows

1. Pull top lip over teeth
2. Push bottom jaw forward and pull bottom lip over teeth
3. Pull up sides of nose, cheeks
4. Lift eyebrows (if you want an extra pull)

- Hold continuously for approximately1minute

Note:
This is difficult and may take you a while to master, start with holding for less time and work up to one minute.

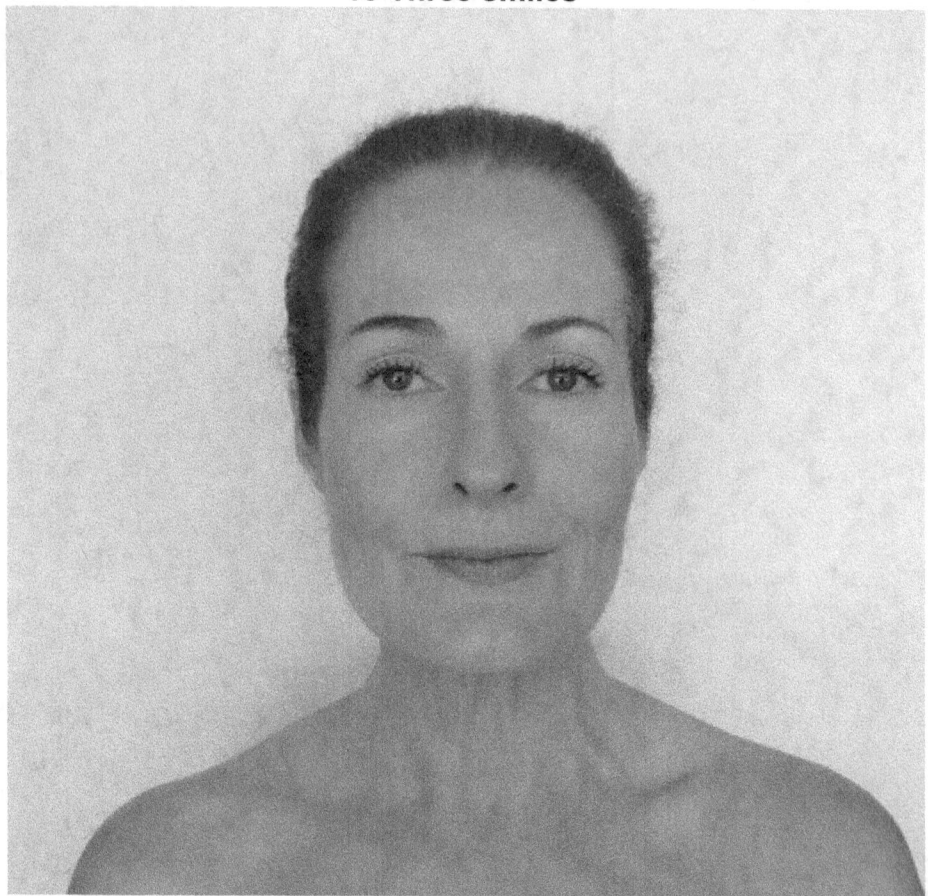

TARGET Cheeks & jowls

1. Lips together smile horizontally, hold for a count of 10
2. Lips together smile more towards the earlobes, hold 10 count
3. Lips together make a huge smile temples, count to 10
4. Lips together smile towards earlobes, hold count to 10

- Keeping the tension repeat

Note:
This is difficult to master. It is like smiling up once, twice, three times then letting go of the smile once, twice and starting up again.

ADVANCED
11 Jaw lifts

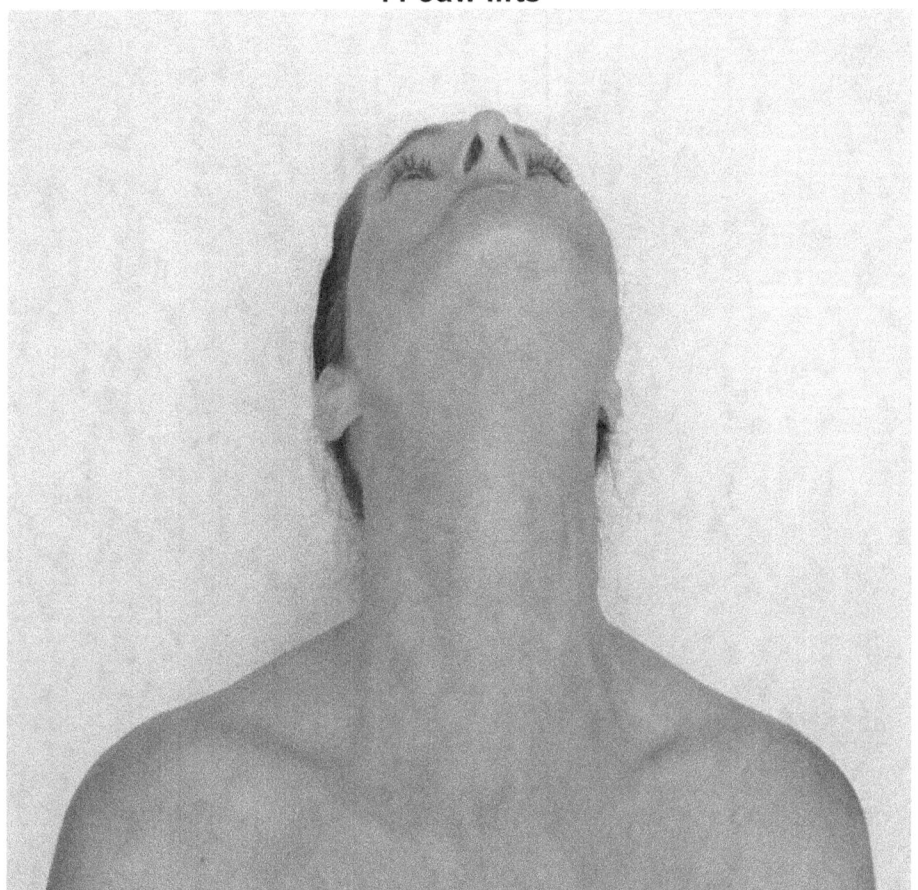

TARGET Neck & chin

1. With mouth closed thrust your lower jaw forwards
2. Make a wide grin curling the lips and corner of the mouth up
3. Extend your neck, relax shoulders and gently tilt your head back
4. With your head resting backwards thrust your chin towards the ceiling
5. Pause for a moment
6. Repeat 10 times in succession
7. Relax
8. Bring head forward

- Repeat

ADVANCED
12 Chin dimples

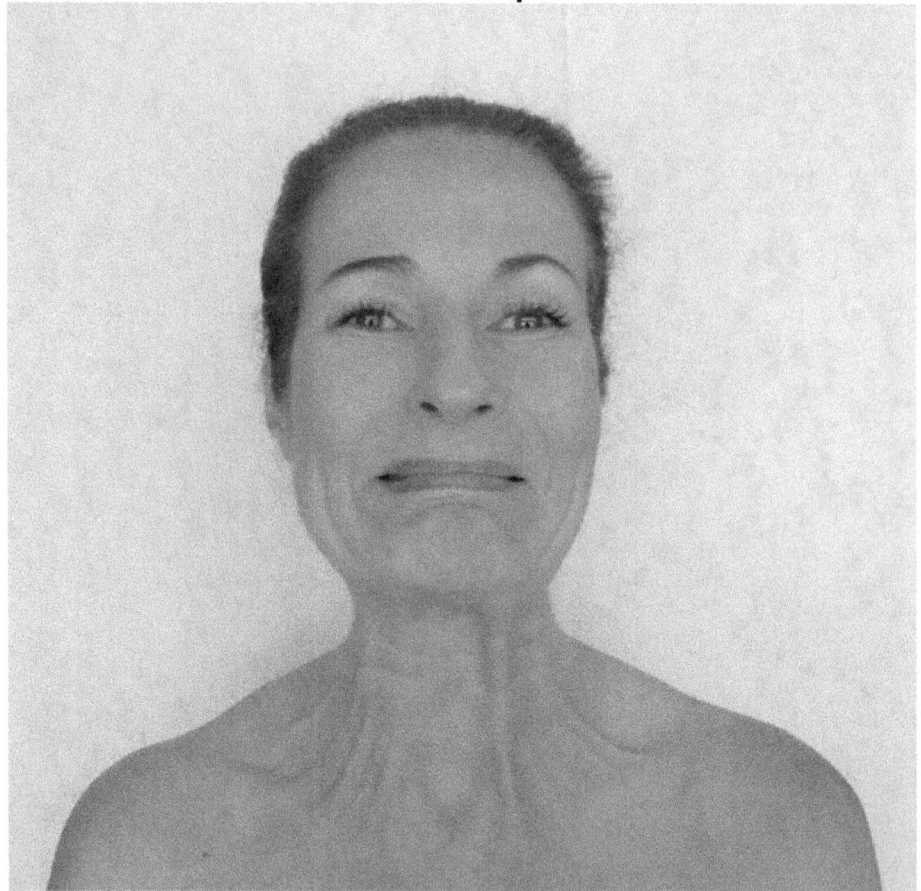

TARGET Chin & neck

1. Close your mouth and clench your teeth
2. Pull the corners of your mouth down, your neck muscles should stand out
3. Force the corners of your mouth upwards, this should make divots in your chin
4. Hold for a count of 30

- Repeat

Note:
This is very difficult, do not strain too much. If you find it too difficult do not repeat.

ADVANCED
13 Neck pull

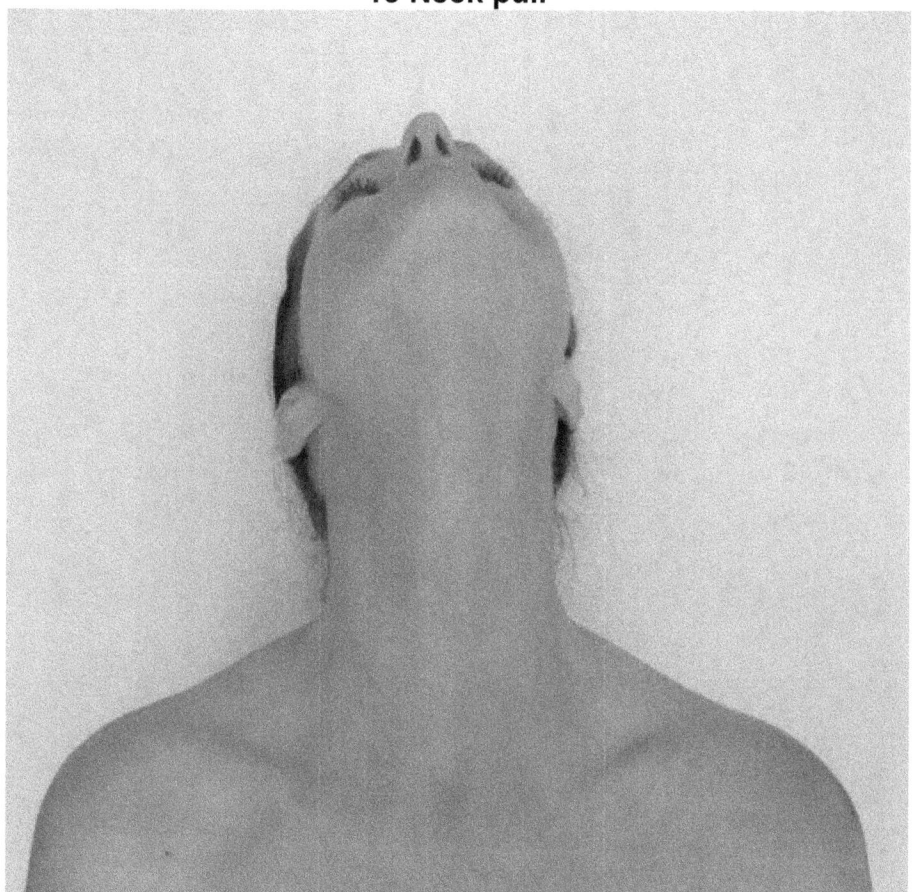

TARGET Chin & neck

1. Sit upright , relax your shoulders , extend your neck and gently drop the head backwards
2. Look at the ceiling while keeping your lips closed and relaxed
3. Open your mouth and bring your lower jaw forward and upward trying to get your lower teeth to cover your top teeth.
4. Hold for a count of 10
5. Relax, bring head gently towards

• Repeat 3 times

Note:
This is hard exercise, use extreme caution.

ADVANCED
14 Neck flex

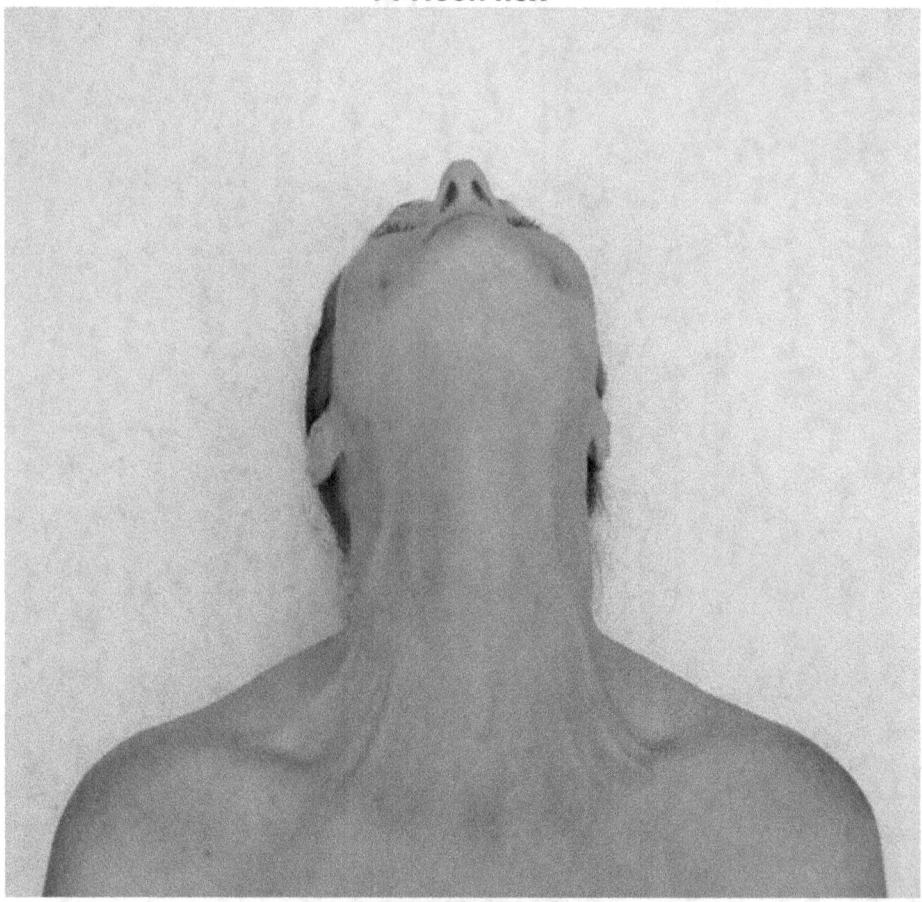

TARGET Neck

1. Clench your teeth and spread your bottom lip,(your neck muscles will flex)
2. Keeping the tension, extend your neck, relax your shoulders and gently drop your head backwards to rest on your shoulders
3. Lift and lower your head 5 times in succession
4. Relax the tension and raise your head to an upright position

- Repeat

Note:
This is a very hard neck muscle exercise, be careful

ADVANCED
15 Ears relaxation

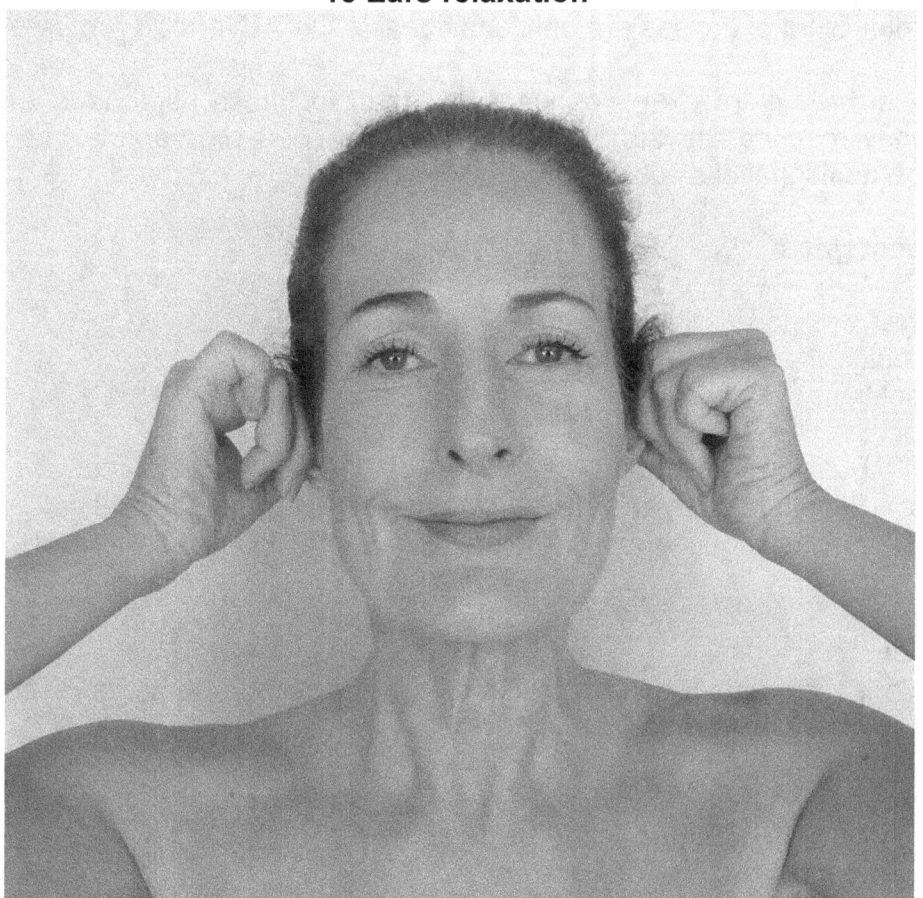

TARGET Ears & jowls

1. Pinch your earlobes and massage between fingers moving up the back of the ear coming all the way to the front
2. Put your index finger in your ear hole and gently rotate your finger to massage inner ear whilst your thumb gently massages the back of the outer ear

Extra Exercises

(Note: Extra exercises are only additional to the **Advanced** programme)

These three extra exercises were designed for those persons who have already reached the target level of the Advanced programme and want some additional choices to attack a problem area.

Widest grins

Target	Upper cheeks
Difficulty	Very hard
Repetitions	30

1. Pull your widest smile, showing your teeth and crow's feet
2. Simultaneously raise eyebrows and cheeks, pause, release
3. Repeat continuously (maintaining the widest smile), slowly, with control

Note:
Smile is maintained, nose wrinkles as the opposite force as the eyebrows and cheeks are worked up and down.

Under eye lifts

Target	Under eye bags
Difficulty	Very hard
Hold	60 seconds

1. Raise the eyebrows
2. Lift the area under the eye, until the lower eyelid is horizontally across the eye pupil
3. Hold for I minute

Side chin thrust

Target	Jowls chin and neck
Difficulty	Very hard
Hold	10 counts
Repetitions	3 to each side

1. Extend your neck
2. Gently drop your head backwards
3. Push jaw forwards
4. Move jaw to the left and hold, for a count of ten. Automatically the left corner of your mouth should rise up to the outer corner of your eye
5. Move jaw to the right and hold, for a count of ten. Automatically the right corner of your mouth should rise up to the outer corner of your eye

Note:
To increase the pull push your lips forwards.

Contact

Visit us
Email
Guestbook
Customer Service
Where to buy Fitface
www.Fitfacetoning.com

Fitface Fusion – products
Coming soon
www.Fitface.co

Follow on Twitter
www.twitter.com/Fitfacetoning

Read our blog
www.blog .Fitfacetoning.com

Watch us on YouTube
www.youtube.com/user/Fitfacetoning
Fitface fun exercises
See Charlotte on ABC News at 5pm WPBF
Fitface professional instructor training
Fitface seminar

Listen to clips on www.Fitfacetoning.com
BBC Radio Solent
Katie Martin